The og Book

Sues pacis in mundum solvite ut bona opera factum eant!

(Let loose the hogs of peace upon the world to go and do their good works!)

The Hog Book

WILLIAM HEDGEPETH

Drawings by John Findley
Photographs by Al Clayton

Brown Thrasher Books

The University of Georgia Press
Athens & London

Published in 1998 as a Brown Thrasher Book
by the University of Georgia Press
Athens, Georgia 30602
Prefatory remarks ("Forward!") to the Brown Thrasher Edition
© 1998 by William B. Hedgepeth
Text © 1978 by William B. Hedgepeth
Drawings © 1978 by John Findley
Photographs © 1978 by Al Clayton
All rights reserved
Printed and bound by Edwards Brothers, Inc.

The paper in this book meets the guidelines for permanence
and durability of the Committee on Production Guidelines
for Book Longevity of the Council on Library Resources.

Printed in the United States of America

02 01 00 99 98   P   5 4 3 2 1

Library of Congress Cataloging in Publication Data

Hedgepeth, William.
The hog book / William Hedgepeth ; drawings by John Findley ;
photographs by Al Clayton. —Brown Thrasher ed.
p.   cm.
"Brown Thrasher books."
Originally published: 1st ed. Garden City, N.Y. : Doubleday, 1978.
Includes bibliographical references (p.   ).
ISBN 0-8203-2018-8 (pbk. : alk. paper)
1. Swine. 2. Swine—Miscellanea I. Title.
SF395.4.H42   1998
636.4—dc21   97-52704

British Library Cataloging in Publication Data available

Lines from Ovid, *Metamorphoses*, translated by Rolfe Humphries,
© 1955 by Indiana University Press, Bloomington. Reprinted by
permission of the publisher.

Lines from "The Second Coming" by William Butler Yeats reprinted
with the permission of Simon & Schuster from *The Collected Works of
W. B. Yeats, Volume 1: The Poems*, revised and edited by Richard R.
Finneran. © 1924 by Macmillan Publishing Company, renewed 1952
by Bertha Georgie Yeats.

Lyrics from "White Freightliner Blues" by Townes Van Zandt
© 1974 by Columbine Music, Inc./ASCAP. Reprinted by permission
of Jeanene Van Zandt.

The Hog Book was originally published in 1978 by Dolphin Books/
Doubleday & Company, Inc.

DEDICATED . . .
to the
millions of
porkers who've
gone to
their final
resting sites
inside us,
and to the
ghosts of
still billions
more pigs
who've
long since
passed away
down the throat
of time.

I'd like to
call them all
by name,
but the list
is long
and I cannot
remember.

                    WBH

Contents

Forward!

On behalf of the 770.4 million hogs around the world—and the uncounted millions more living deep in forests and uncharted reaches who managed to escape the official census—I bid you welcome to the return of *The Hog Book*. This revolutionary opus was originally published in 1978 and was an amazing best-seller all over the world—translated into Chinese, German, Spanish, and other languages—until it was abruptly taken out of print, along with other books, as a result of a decision by the U.S. Supreme Court (see the *Thor* decision, 1979). But now, thanks to the wisdom and visionary spirit of the University of Georgia Press, you are able actually to hold this newly-minted and updated volume in your hands, to page through it in a spirit of growing wonder, to share it with friends and to keep it in some special spot safe from the prying eyes of the judicial system.

The Hog Book is simultaneously as serious and silly as the hog himself, in all of his earthy nobility and churning porcine vitality. At the time of the book's original publication, our society was clearly overdue for a new consciousness about these creatures. To a certain extent, that has been achieved in the intervening years. Pet pigs, for example, are no longer uncommon. There are now an American Miniature Pig Association and a North American Potbellied Pig Association, which together have a registry of over 27,000 pigs kept as household pets. Hogs are also used by various police departments to sniff out drugs and contraband. There are racehogs and watchhogs and hogs trained for combat duty—all foretold in my chapter "The Porcine Potential."

But to me, what is most astounding (also as predicted) are the cascading breakthroughs on the medical front. It has long been known that the hog's heart and cardiovascular system are more like man's than any other creature in the known universe. For the past twenty-five years, hog heart valves have been used to replace damaged human heart valves. John Wayne had them, as did Lewis Grizzard. But in clinical trials starting this year, entire hog hearts are to be transplanted into the chests of dying children. Moreover, a handful of laboratories in England and the U.S. are presently injecting human DNA into hog embryos, raising the hogs in germfree conditions, and using them as organ transplant donors. In the future, if you should find yourself on a long waiting list in need of a new heart, lungs, kidney, or a liver, you can either accept such an organ from a genetically altered pig specially bred to be a donor or you can continue to wait until some hapless soul falls off his Harley-Davidson. And God only knows where that guy's been.

As if interchangeable organs were not enough, consider that in 1997 a research team near Boston figured out how to take immature animal cells and inject them directly into the brains of patients suffering from Huntington's disease or Parkinson's disease, both of which gradually destroy the brain cells of their victims. The doctors put these healthy new cells into the patients' brains, and, as they develop and grow, they take over where the diseased cells left off. This is the first time EVER that tissue from another animal species has been transplanted into the brain of a human. And what kind of animal do you suppose it is whose cells the doctors deem worthy of placing into human brains? Pigs, that's who! Among the dozen patients who have undergone what is called "pig cell surgery," the resulting improvement is said to be "simply miraculous."

The upshot of all this is that in increasing numbers there are going to be people among us who are part pig. They're already out there. You may be one yourself. And since you won't ever be able to know exactly who is part pig and who isn't, common decency might suggest that you begin today to start rooting out all the thoughtless and unkind expressions regarding pigs, hogs, or swine in general that are so loosely and hatefully salted throughout our language. This can be your own little contribution to the coming shake-up in the social role of hogs and to society's deeper appreciation of essential hogritude.

Reading *The Hog Book* is another productive step in this direction. Still, as comprehensive as this book aspires to be, there remain swirling mysteries unanswered, perhaps unanswerable. A major mystery came from a small story I read a few years back in the *New York Times* wherein two adult hogs were spotted as they trotted alone down the streets of lower Manhattan in the pre-dawn hours. Police were unsuccessful in locating them until, later that day, these same two hogs turned up on Staten Island! Now how in the world did they do that? Hogs can swim well enough, but that would have been stupid. Clearly they got aboard the Staten Island Ferry, but how? Disguised as what? And where did they get the ticket fare? Merely contemplating such a thing can lead to madness.

Meanwhile, I am profoundly grateful (and the majority of hogs are certainly grateful) to all the major talents who had a hand in the creation of this book, especially to my principal collaborators, Al Clayton and John Findley, in addition to such poets and creative contributors as Dr. John Hatcher, Coleman Barks, Roy Blount, Jr., Townes Van Zandt, Frank Trippett, Michael Catalano, Ralph McGill, Jr., Dr. Bonnie Fritz, and the multigifted Grace Zabriskie: each of whom has solemnly foresworn the eating of pork.

So now, my friend, it's time to leap in and begin to wallow, bearing in mind that you will eventually emerge as a changed and better person. It begins with a journey. And ends with something you may wish forever to carry in your wallet as a badge of your enlightened heart. Root on!

William Hedgepeth

January 1998
Tangier, Morocco

The Hog Book

I
The Hog Mystique

And there in a wood a piggy-wig stood
With a ring at the end of his nose . . .
Dear pig are you willing
to sell for one shilling
Your ring? Said the piggy, "I will."
—Edward Lear

"Hoover?"

". . . thirty-four, thirty-five, thirty-six, thirty-seven, thirty-eight, thirty-nine, forty . . ."

"Hoover, listen, man, there really . . ."

". . . forty-one, forty-two, forty-three, forty-four . . ."

". . . really must be some . . ."

". . . forty-five—*quiet!*—forty-six, forty-seven . . ."

". . . better way of doin' this whole thaing."

". . . forty-eight, forty-nine, fifty, fifty-one—goddammit—fifty-two —shut-up—fifty-three . . ."

"Hoover, I don't like this a-tall."

". . . fifty-four, fifty-five—shuttupdammit—fifty-six . . ."

"On top of everything else, they'll prob'ly . . ."

". . . fifty-seven, fifty-eight, fifty-nine, sixty . . ."

". . . puke."

". . . sixty-one, sixty-two, sixty-three, sixty-four . . ."

"First big bump we hit, and ever' single gawddamn one of 'em— *Bllaaaaahhhhppppp!*—puke all at once. Hoo boy!"

". . . sixty-five, sixty-six—dammit, Doyle, shut *up*—sixty-seven . . ."

Hog, hog, hog after hog: At a perfect steady pace the column calmly lumbered forward like a self-willed stream of meat, flowing up the inclined rampway from the pen to the edge of the truckbed, each hog filing along in an orderly manner while at the same time sharing with each and all of its co-hogs a mildly euphoric eagerness. Some of the hogs paused now and again, whenever the procession slowed, to nuzzle gently with their immediate neighbors . . . and only a rare one or two along the line occasionally turned their heads to gaze back at the place, the home, they were leaving.

"Hoover?"

". . . sixty-eight, sixty-nine—*shush!*—seventy . . ."

"Hoover, you know you're gonna be the one who's gonna . . ."

". . . seventy-one, seventy-two, seventy-three . . ."

". . . hafta clean up my truck if they do."

". . . sevent—GODDAMN YOU, SHUT—seventy-two . . . sev . . . *ahhggg.*" At the instant he raised his voice the eight, ten or so hogs toward the top of the ramp had jerked back in fright and then made an abrupt lurch forward, spilling wildly into the truckbed—with all the rest behind them following close on their heels in the same spirit of maniacal abandon. "Oh for Christ A-mighty sakes," howled Hoover, slamming at the truckside with his big fist, "ya made me lose count."

Back in the pen, picking up the message from the others, the remaining porkers began grouping together at the base of the ramp, then funneling themselves briskly up, up the wooden rampway—with the accumulated tappings of their trotters swelling into a brittle staccato clatter—and then on into the covered truckbed to become part of the moil of massed pig inside, all grunting, snorting and idly rooting at the floorboards.

And then the older man, the farmer, stepped forth from somewhere, in his mud-caked hightop shoes and too-big overalls, closed the gate at the end of the ramp and drawled in his reedy twang, "They wuz eighty-seven of 'em." And a pause. "S'posed to a-been, anyway." Then he inhaled deeply, gave a brief, barely audible sigh and squinted in the direction of the two younger men. "I reckon that's about it." And on saying this he dropped off into a silence and stood frozen for a moment, gazing off from between his half-squinched eyelids, through thick specs and then all the long way down a lean hatchet nose, staring absently past a pasture. Rufus Hoover, his nephew, just pursed his lips and said nothing; and Doyle Moon picked up the cue and fell quiet too, partly since neither of them had any idea what they ought to say. But then the old farmer spit, shrugged and offered a bland smile. "You do what you do," he nodded. "Land gets corned out, cash crops wind up ruint, et up or overrun with chinch bugs an' childs of the earth an' beetles an' hoppers an' thaings that suck plant life to death. Yessir . . . yessir, you do what you gotta do. Shoulda kept these hawgs another month or two at least."

The sun by now had gone nearly down, and the congregation of hogs aboard the truck was already making shifting, restive, twilight noises as Moon and Hoover scouted around the truck securing all the latches and checking the gauges, the tires, the fluids and other mechanical variables. Then they climbed into the cab with the air of pilots about to be launched from an aircraft carrier and scrooched themselves down into the most nearly comfortable contortions each could manage in anticipation for what they both knew would be a long haul across the night.

Once behind the wheel, Moon, an independent trucker by trade—stockily small, slightly gat-toothed, affable and twinkle-eyed—began to feel back in his own element at last. "Hoooo weee, man," he said, snapping the chinstraps of the stars-and-stripes Captain America motorcycle helmet he wore whenever he felt like looking official, "I swear there must sure be some better way o' doin' this whole thaing. I don't know nothin' about haulin' hogs, but I know *that*." Now he tucked his bell-bottom trouser cuffs into the tops of his cowboy boots and tested the brake. Each time he hit the pedal, the glass eyes of a fluffy white cat reclining on the right corner of the dashboard flashed bright red. "*Faaan*-tastic," he said, as if to declare authoritatively that every last little thing, including even this, was in good working order. Then he fired a quick devilish wet leer toward Hoover and reached for the ignition switch.

The diesel awoke—*WHooom-blapblapblapblap-Blaaaroommm*—with a gruff splutter and simmered down low into a drawn-out growl which Moon mused over for a moment before finally shifting into gear. "*Faaan*-tastic," he muttered to himself, satisfied.

With the first snort of engine, the hog cargo hushed and remained quiet except for the few jostling grunts and stumblings they made among themselves as the big truck began to lunge gradually forward across the stretch of gravel drive that led to the highway. Then bump-*bump* in front, a *bump-bump* in back, and all the tires, now over the asphalt edge, rolled with a smooth, clean whine slowly picking up to the pitch of road speed, whipping a brief pall of dust rearward.

Out over the pancake countryside of central Florida, the flatly clouded skies of early evening loomed with especial wideness. Hoover blinked, almost audibly. He sat like an indifferently hunched hulk, with his large hands between his knees and his neck drawn into his shoulders, and looked out at the road ahead. "Baa baa boo dee," he began to sing to himself, absent-mindedly recalling some jivy tune. "Dooby wah bee doo," he snapped his fingers and shook his head with a faint rhythm from side to side and up and down in lumbering little jerks like a young moose fascinated with the weight of his antlers. Hoover had the classically guileless face of an Iowa farm boy, slightly aged and at once ecstatic and vacant. His chipmunk-colored hair was just barely beginning to thin on top. But hardly anyone could ever see the top anyhow on account of his size. He stood over 6'5" in a heavy-set, big-boned body and was altogether infused with the innocent confidence of someone so ominously large that no one in his right mind would ever dare try to mess with him. Nobody ever had. And so, unlike the other good old country boys who always seemed to feel compelled to contest with one another over issues of manliness, he'd always felt free to concentrate pretty much entirely on the more simple, peaceful things he enjoyed, like anything to do with farming and—most especially—hogs.

Then Moon, who spoke in quick bursts and with never much seriousness, said, "Listen here, sugartit, I hope you know this ain't my kinda thaing a-tall. They tell me it's bad luck to truck a hog. Most particularly in the dark. You just never know what they're doin'." He smirked like a sleazy cherub and cut his eyes in Hoover's direction. "Hee."

"Well," said Hoover, making circular motions with hands whose gestures rarely seemed to bear much relation to whatever he happened to be saying, "a' course hawgs'll shrink when you truck 'em in the daytime— about a pound an hour. They can overheat too. An' some of 'em die. It's just better at night."

"No shit!" said Moon, whose thoughts had already flitted off somewhere else.

Hoover, who had recently forsworn chewing tobacco, now proceeded with great painstaking care to roll some Bull Durham into a vaguely shaped cigarette, then lit up, put on his wire-rim specs and fell back silent into his heaped hulk position, gazing out the side window.

The road skirled along in the muted autumnal light, past paintless shacks with gnarled TV antennae; past feeble barns and row crops and Jesus signs and drive-in movies billing double-featured cycle flicks; past backwater chiropractic clinics; signs for Sister So-and-so, palmist and adviser; men chugging home on their tractors; and, here and there, past a ratty-haired sawmill gal trudging along the grassy shoulder.

"Wellsir, I just tell ya this," Moon nooded after a long pause, "I sure as hell don't like to turn *my* back on them critters. It's like they say, hogs move in mys*teri*ous ways."

"Toads are what get me," Hoover said, and began to grimace a little at the thought. "Toads are treacherous. Hell, I heard of this one fella out west—guy owned a big ranch and all—an' one day he just upped an' quit everything else an' went into toad farming full-time. Nobody knew why. After a while word got out that he was bein' paid, actually paid by the CIA, or somebody like that, to raise these goddamn killer toads. Taught 'em to carry poison or somethin'. Just gives me cold chills to think about." Hoover took a long drag on his cigarette and shook his head slowly as he exhaled a thin stream of smoke.

"Critters'll get ya, man," vowed Moon, with his eyes now carefully fixed on the darkening highway. "One way or 'nother they'll get to ya." He leaned forward and flicked on the headlights.

In the drowsing silence behind the cab, Moon's and Hoover's voices grew muted into a softly fragmented murmuration of words, the way it sounds to a small child waking in darkness to hear his parents making gentle talk-sounds with no seeming purpose or meaning. Here in this brooding quiet—broken only by sporadic snorts and muffled grunts—lay outspread a protoplasmic sea of fleshy backs in swells and eddies: tightly jammed together hogflesh . . . dulled and shifting in the shadows . . . out of focus . . . unresolved.

❁   ❁   ❁

An aura of mystery flows forth in the face of the awesome simplicity and eternalness of the common hog. In terms of basic animal design, the hog is a primary form. Of all domestic mammals, he is the most functionally shaped and the most ancient, having undergone less evolutionary alteration of his physique than any other farm creature—less even than the average farmer, as a distinct species of human.

So far as scientists, anthropologists and historians are able to determine, pigs have been rooting about the globe's surface for at least 45 million years now, whereas men emerged and commenced to root around only about a million years ago, a mere gnat of time as such things are reckoned on the hogspan scale. And in all the ensuing aeons, hogs have remained structurally intact: the same skull and skeletal pattern, the same fourteen ribs and basic bone-fittings, the same forty-four teeth and seventy-five feet of intestine, the same cloven hoofs, the same sweet snout, the same apparent passions and ardent desires. In fact, were the very first of all earthly porkers—the *Ur*-hog, the proto-pig, the primal, archetypal, first-off-the-assembly-line swine—to appear suddenly, and without comment, upon a city street he would be instantly recognizable as just that very thing. The fundamental essence of hog flows up intact from the bottomless wellsprings of the past. Hogness, then, is a constant; "Hog" is an Absolute, unchanged, *a priori*. "Hog" connotes something seemingly everlasting and serene—and hence he tends to be agonizingly uncomfortable to man, both as a concept and, God knows, as a ubiquitous creature with which to cohabit the earth. A creature of which very little is really known, other than the fact that he persistently exists.

Humans, on the whole, perceive their own innate uneasiness about this, not only in the actual presence of the living hog-on-the-hoof but also purely upon reflection in the solitary darks of their rooms, or wherever else the occasion might arise that they should try to wrap their minds around the thought of Hog. It has ever been thus. Among many peoples throughout history the proposition of "Pig" has evoked such an enormity of deep-rooted apprehensions that merely to voice the creature's name was—and still is in some places—anathema and a thing to be left unmentioned at all costs. In Scotland, for example, traditional belief held that the simple enunciation of "pig" was sufficient to bring on disaster—particularly so if the word happened to slip out when at sea. In various other areas of the world, the verbalization of feared, myth-shrouded names such as ——* is side-stepped by using some euphemism with a meaning adequately benign so as not to ruffle malevolent spirits. Thus, among certain cultures, pig is not "pig" but is rather the "short-legged" or "the grunting animal" or "the beautiful one" or even, in parts of China, "the long-nosed general."[1]

Nevertheless, even among men who are not caught up in the extremes of "cultural hogrophobia"—which, according to Dr. H. R. Datrum, an animal physiologist who did the pioneering work with turkeys, may be defined as "a socially institutionalized fear of hogness"—the hog hangs on the horizon as an open-ended enigma. Admittedly not often, but now and again, at certain cocktail parties and at Confession, over lunch counters, deep in Kentucky coal mines, at fashionable ski resorts, on the *Champs Élysées,* and on trolleys, trains, planes and frequently even in barbershops, one runs into (or perhaps one asks) that ultimate riddle: "Why hog?" To which the reply most usually returned—if it is answered at all and not met with a diffident shrug—the reply which has by now become the standardized admission of man's intellectual impotence when he's forced to grips with the subject, is: "Verily, hog *is* because Mother Nature in her infinite wisdom chose to have it thusly so. That's why."

Hence, the mind must shrink back from such cosmic probings, must resign itself to lesser questions and lesser answers, or to traditional articles of faith clung onto in the absence of a generally accepted rational riposte. Take, for example, the still open question of hog origin. Lacking anything better, the unspoken general contention has always been that hogs are not supposed to have *come* from anywhere; rather they just *are,*

* Pig.

like feminine laughter and hangnails and morning dew. Yet deep within that innermost center-cut loin-eye of the human heart there dwells the need to know. Toward this end, there exists, in some, the willingness to start *a posteriori*, if need be, and build up the body of empirical fact beginning from the level of the humblest trough and ascending to whatever awesome height the search may lead. And it begins with that simple yet formidable inquiry: "What (or who) is the hog?"

Formulating the fullest and most "objective" answer attainable must start with assimilation of cold fact. We gather about us those things we "know," or, more precisely, those items allegedly "known" by certain individuals who themselves are acknowledged to be in possession of some specific body of data which—assuming that the most scrupulous allowances have been made for (1) possible alterations in the subject matter; for (2) acts of God; and for (3) the tenebrous nature of language and all the subtle nuances inevitably lost in the act of translation—comprise some portion of the sum total of human observations concerning those qualities peculiar to the creature that is referred to, by some, as a "Hog."

Viewing "Hog" in the proper perspective requires, at the outset, a perusal of the accepted family tree:

> CLASS: *Mammalia:* animals possessing teats for suckling their young.
> ORDER: *Artiodactyla:* animals with an even number of toes—which also includes the deer, giraffe, camel, cow and hippopotamus.
> SUBORDER: *Suiformes:* animals that have incisor teeth in both jaws plus canine teeth that are enlarged and extended into tusks (or tushes).
> FAMILY: *Suidae:* animals of moderate size, covered with bristles and having an elongated head that ends in a snout. They are omnivorous, terrestrial, and they tend to grunt.
> SUBFAMILY: *Suinae:* (something I shall have to remember to look up before this goes to the printers).
> GENUS: *Sus:* animals including the Wild Boar of Europe and of India, as well as all breeds of domestic pigs.

Yet even taking all these described qualities into account, there remain great gaps and questions unanswered for anyone seriously seeking to comprehend that which is distinctively "Hog" *vis-à-vis* other animals. Now there are, to be sure, profound similarities with all manner of other creatures; but what ultimately manifests itself through the painstaking process of comparison are those even *more* profound points of contrast which serve further to depict both the physical and behavioral aspects of "Hog."

How to tell a hog from a kangaroo: Your basic kangaroo is of the family *Macropodidae*. Like the average hog, the Great Gray Kangaroo (species: *Macropus gigantus*) weighs over two hundred pounds. Similarly, he has mildly elongated ears, dark eyes and a long, tapering face terminating in a nose. Both kangaroos and hogs are social animals and are both mistrustful and easily frightened. They differ, chiefly, in their means of perambulation. Whereas a hog will stroll, saunter, trot and even gallop (always using all four feet), the kangaroo moves rapidly by *bounding*, often as much as eleven yards at a jump. The kangaroo gets about on two hugely outsized back legs which, when used for more leisurely movement, give a group of kangaroos the initial appearance of odd-looking small-headed creatures scooting about in wheelchairs. Now as it turns out, hog's singularity emerges here by way of the *absence* of these things. A hog can be distinguished from a kangaroo even further by virtue of the fact that it is more difficult to coax a hog to take up the art of boxing. Moreover, sows do not have pouches, wouldn't use them if they did, and manage to produce offspring referred to as "piglets" or (after weaning) "shoats," but never "joeys."

How to tell a hog from a seal: The "true seal" (a member of the *Phocidae* family) physically resembles the true hog in terms of their both being nearly neckless, relatively short-haired and insulated with layers of subcutaneous fat. Both also grunt, both swim and both are demonstrably intelligent. The hog's uniqueness in comparison with the seal springs forth with respect to his *not* having paddle flippers, *not* being earless and *not* running in "schools." Another significant definitive factor is that whereas the seal's chief natural enemy is the Grampus or killer whale, the hog's is not.

How to tell a hog from a bear: This is a hard one. The bear (family: *Ursidae*) has a proportionately short tail, a pointed muzzle, small eyes, erect ears, forty-two teeth and a heavy body on short, strong legs—all, obviously, quite similar to the hog. Pandas, in fact, with their white midsections contrasting with predominantly black bodies, display very nearly the same markings as purebred Hampshires. The subtle quality of hogritude is revealed, in this instance, by taking account of the fact that hogs do not choose to walk on their hind legs; that hogs are too innately curious and active to afford themselves the luxury of hibernation; and that hogs, unlike, for instance, Polar bears, simply could not thrive or even stay alive on drifting ice floes, glaciers and all such stuff as that.

Nonetheless, though certain markings may be the same, the hog stands out as *hog* because his is a coat of neat, short bristles, while the bear is just all-over hairy. The little verse that follows has been found quite valuable as a mnemonic device for defining in people's minds the elusive borderline between bear and hog as distinct life forms:

> *Once I stopped awhile to ponder*
> *As I chanced upon a bear;*
> *Quite unlike the tidy pig, his*
> *Fangs are long and so's his hair.*
> *(Pick a pail of collard greens to*
> *Store in Mama's Frigidaire.)*

HOW TO TELL A HOG FROM A POSSUM: The common opossum, though part of the *Didelphidae* family, is very like a hog in his wide geographical distribution, his omnivorous diet, his tapering snout and his rate of bringing forth offspring. Hogs, however, come into their own in comparison with the possum, by virtue of their having a much more restrained and aesthetic tail, tastefully tufted and not all ratty-looking. Then too, a possum, if threatened, will feign death, going limp with eyes closed and tongue lolling out, while in no recorded instance has a hog ever stooped to fakery of this sort.† The most recent research into possums sheds light on a further sole characteristic of hogs. Possums are most readily found on the edge of roads and highways, dead. Their quaint knack for running into cars and trucks has traditionally been ascribed to poor eyesight. An emerging new theory suggests that it is actually an instinctual *goal* of the possum organism to get run over by something—there being very little else to do. High-speed cameras on the bumpers of pick-up trucks have documented photographic evidence of possums lurking in high grass waiting for the onrush of a speeding vehicle. Separate studies of countenances of the subsequent corpses reveal, almost to a possum, facial expressions variously described as "benign," "blissful," "euphoric" and "fulfilled." The essential point here, you see, is that hogs don't do this. Rarely does a hog get hit by a car, and never does he contrive to do it on purpose. As a side point: Owing to such features as pigbulk plus a low center of gravity, it scarcely happens that an auto or its driver emerges unscathed from a high-speed encounter on the highway with a hog.

† A number of these hog/possum dissimilarities don't necessarily hold true when it comes to contrasting the hog to the terrestrial possum (genus: *Peramys*) of Brazil and Argentina. This particular possum has a short tail, a more straightforward manner, a heavier body, and—taking all pertinent qualities into account, as well as the views of one or two authorities—is himself, in fact, actually (and from all the evidence, rather obviously) a hog.

How to tell a hog from a crow:
The Common Crow, of the *Corvidae*
family, shares with the hog a fondness
for grubs and, in wintertime at least, an
omnivorous diet. Secondly, crows, like Poland
China and Berkshire hogs, are black. And
finally, both crows and hogs can roost in trees
—a fact recorded by a foreign dignitary's visit
to a flood-prone portion of southern Indiana in
1843, where he wrote: "This high water is said
often to present an interesting scene. Hogs . . .
have been found on low trees, where they had
sought protection."[2] The unique features that
single out the hog begin with voice. Depending
on the species, the crow may go either "caw" or
"car," whereas the hog commands a whole array
of vocal sounds on a somewhat lower register.
The chief distinguishing characteristic, however,
lies in the legs. The singularity of the hog can be
ascertained here not only by noting that hog is in
command of exactly *twice* the number of legs as crow
but also by observing, in a side-by-side examination,
that the legs affixed to the hog are muscular and
cloven-hoofed, as opposed to being brittle, spindly
stalks that branch out into clasping claws. The
crow, too, falls short of the hog in terms of his
general desirability as an item of human consumption. Though there *are* people who relish
crowmeat above almost all else, these same individuals willingly acknowledge that the entire
adult crow, beak to claw, can be devoured with
little discomfort at one sitting—a phenomenon
which can be measured against the much
greater time required for a single man to
consume one (1) complete hog.

Generally, then, there unfolds from all this a nebulous picture of behavioral characteristics that supplements your cold-eyed academic descriptions. Everything included and implied within the vast concept of hogosity is, after all, far more weighty and chimerical than any mere verbal definition could ever possibly encompass.

Here, for example, is the Encyclopaedia Britannica's attempt: "SWINE, a name applied to the domestic pig, but also used to include its wild relatives. The swine are found only in the Old World. They are characterized by elongated head and mobile snout, with an expanded, truncate, terminal surface in which the nostrils open; narrow feet with four toes, the outer pair not reaching the ground when walking. The canines in both jaws curve upward, forming large tusks, best developed in boars. The hair is coarse and bristly, while the skin is thick, underlaid by thick fat. The tail is moderately long, generally tufted."

Britannica does a bit better with "pig": "PIG, a word of obscure origin . . . a common name given to domestic swine of any age. Hog is used collectively with much the same meaning as pig. Swine is applied to any hoofed animal of the family Suidae . . . Pigs are rotund-bodied, short-legged artiodactyl animals of omnivorous habits, having thick skin from which grow short, coarse bristles, a long mobile snout, small tail and feet with two functional and two non-functional digits. A mature pig has forty-four teeth, carries its head low, and eats, drinks and breathes close to the ground."

All of this is quite true, of course. But what says more than anything else is the Britannica's single sentence later on under the same entry: "The origin of the pig is shrouded in *mystery*."

Ah, we rail and rage and caterwaul and carry on; we storm through the night, over broad oceans and narrow bars, across silly bridge tables and smug classroom rostrums, with hype and pap and vapid drivel, with legislation, propaganda and songs of total self-assurance saying to ourselves once more that all the wonders of the world are finally rendered impotent and limp within our mental grasp. And yet here's this thing. This hog. This hulk of deceptive simplicity: wraith, myth, phantom, shadow of forgotten ancestors . . . a fabulous labyrinthian being swathed in fog, arcanal and everlasting . . . a rude surface enveloping an enchanted loom. Even those who deal with them day by day willingly confirm their own ignorance—tinged with a constant awe at the hog's complex inner workings . . . and at the enormity of that which remains to be known. As Blake Pullen, former officer of the Georgia Swine Growers Association, lifetime student of the pig and practical hogman *par excellence,* put it: "There's a lot of things to be learned about hogs.

We're in our infancy in knowledge on hogs, in terms of research and everything else. There's not too much good research goin' on on hogs today, and so what we're all doin' is just sorta flammin' around out there in the dark."

*　　*　　*

With darkness it had grown cooler in the covered truckbed, though the living cargo within remained packed together flank-to-flank, cheek-by-jowl with a smoldering and oppressive density. There was minimal movement: just eighty-seven vaguely individuated bodies standing that special way their trotters cause them to stand which makes them look as if they're forever on tiptoes. Some, through the slatted openings of the truck panels, watched the dim countryside rush away in a stream, yet they displayed physical responses no different from any of the rest: simply stood there, swaying slightly, each compressed among and alongside one another like so many shadows whose separate images, at moments, appeared to combine, making all of them, as a whole, look like multifarious facets of one big substance. And at moments, too, their collective breathings fell together in rhythm; and this massed bulk slowly pulsed and heaved like some continuous entity, amorphous, alive and very large . . . resembling something subaquatic, formless, flowing . . . until now, at last, it seemed to congeal itself into a huge lone organism, full-blooded and freshly stepped out of the primeval ooze upon some small expanse of earthscape. Here were hog-forms taking shape . . .

*　　*　　*

". . . and during that time, too, there were some experiments made with genetics," the man explained in a most patient voice, "and we have the results all over the earth today in different forms, among them several forms of monkey life, the orangutan and others—some of the animals that man claims to have descended *from*. Actually, these creatures were the results of genetic experiments in the laboratory where we began by producing life in a human embryo and then confused it genetically in such a way that it came out different. Now, the hog is one of these creatures."

Joseph Myers, a middle-aged professional engineer and former Air Force officer, is presently a resident of Lexington, North Carolina. Before that—in fact, several hundred *lifetimes* before—he dwelt in the land of Lemuria, sometimes called Mu, where a civilization emerged and expired upon a continent that vanished beneath the ocean long before man began to record his history. All of this is said to have taken place in some distantmost niche of the planet's past, hidden beyond the reach of rational intellect. Lemuria is a psychic-historical phenomenon containing elements of science mixed with mysticism, of logic and preposterousness, of the incomprehensible, and of the possible. Quite possibly the hog was made here. Modern science simply offers no other explanation for his presence.

"We may assert," states *The Story of Atlantis and the Lost Lemuria,* published in 1904, "that the outlines of continents and islands have never remained for an hour, nay, even for a minute, exactly the same. Although the lost continent of Atlantis has so far received scant recognition, the general consensus of geologists has for long pointed to the existence, at some prehistoric time, of a vast southern continent to which the name Lemuria has been assigned . . . A number of circumstances suggest that the primeval home of man, the possible cradle of the human race, was a continent now sunk below the surface of the Indian Ocean, which extended along the south of Asia . . . toward the east, as far as the Sunda Islands; toward the west, as far as Madagascar and the southeastern shores of Africa."[3]

From all that the finest psychic sources seem able to discern, Lemuria and Atlantis were co-existent, at least for a while. Lemuria "appears to have existed from early Permian times" (i.e., around 215 million years ago). Atlantis, lying off Africa's northwest coast, became inhabited somewhat later than Lemuria and was ultimately submerged in its entirety "by great tidal waves" in 9564 B.C. Lemuria, however, had "perished by volcanic action before the beginning of the Eocene Age" (sixty or so million years ago).[4]

Now the earliest man-like creature that man himself chooses to recognize is *Pithecanthropus erectus,* the Java Man, who slyly stashed personal collections of his own bones here and there, in such a way as to make us think that he, as First Man, suddenly popped forth from somewhere or other and then upped and died roundabouts of a mere million years ago. This man didn't, of course, choose to leave word along with his bones that right next door to Java, a whole *race* of folks bravely went down with their continent several million years before his own appearance. "The evolution of this Lemurian Race, therefore, constitutes one of the most obscure . . . chapters of man's development, for during this period not only did he achieve true humanity, but his body underwent the greatest physical changes."[5]

Lemurian Man allegedly dwelt and prospered in the Mesozoic Era, between 60 and 190 million years ago. It was during this time that his gelatinous body became solidified and sprouted bones, along with hair, teeth, eyes and other personal effects. Before he acquired these accouterments, the very meaning of the term "Man" was rather up in the air: ". . . they would have appeared to us as gigantic phantoms, if we could have seen them at all, for their bodies were formed of astral matter."[6]

Which led, in time, to this: "The following is a description of a man of one of the later sub-races. 'His stature was gigantic, somewhere between twelve and fifteen feet. His skin was very dark, being of a yellowish brown colour. He had a long lower jaw, a strangely flattened face, eyes small but piercing and set curiously far apart, so that he could see sideways as well as in front, while the eye at the back of the head—on which part of the head no hair, of course, grew—enabled him to see in that direction also. The head sloped backwards and upwards in a rather curious way. The arms and legs were longer in proportion than ours. The hands and feet were enormous, and the heels projected backwards in an ungainly way. The figure was draped in a loose robe of skin, something like a rhinoceros hide, but more scaly. Round his head, on which the hair was quite short, was twisted another piece of skin to which were attached tassels of bright red, blue and other colours. In his left hand he held a sharpened staff . . . In his right hand was the end of a rope . . . by which he led a huge and hideous reptile, somewhat resembling the Plesiosaurus. The Lemurians actually domesticated these creatures . . . The appearance of the man gave an unpleasant sensation, but he was not entirely uncivilized!'"[7]

In time, Lemurians built great cities near what is now Madagascar, as well as "30 miles west of the present Easter Island," and came to represent a "highly evolved humanity."

Joseph Myers nods his head. "I've had," he declares, "some psychic experiences in which I seem to remember living in Lemuria at a time when we had reached a marvelous level of civilization." Myers, an admirer of the late clairvoyant Edgar Cayce and a former member of the Cayce Foundation's Association for Research and Enlightenment at Virginia Beach, Virginia, has traced his reincarnated existence throughout history along a veritable conga line of past human lives. "Each of us," he explains, "is an evolving being that is electrical in its make-up with this physical body gathered around it. So this being is evolving and containing memories of everything that has ever happened to it—but not in the form that the conscious mind can grasp." His own awareness of former lives in previous bodies comes as a result of solitary meditation, in a full lotus position, wherein "you try to stop and hold the conscious mind and anything coming through the five senses until, you might say, you're opening the door to another set of wave-lengths that pour in information from your own mind, but you're unconscious of it happening. Then it filters into your conscious mind in the days and weeks that follow."

Myers, a very open and non-aggressive man, observes that in Lemuria "we had, in many respects, a very beautiful society that moved on beyond to become more technical. The genetics experiments were scientific. They brought into existence plants and animals that had never existed in the earth before. There were different characteristics in matter, if you can imagine that. It was a time when animal and human bodies could intermingle and produce offspring.

"The aim of the experiments was to produce useful animals for servitude; not produce them to eat. I think that the hog, in a sense, was somewhat of an accident. They were experimenting and didn't know what they were going to get. They were trying to develop intelligent animals that could be, in effect, slaves to man. And the hog came out with a body that he has. He was undoubtedly a very intelligent animal and so they preserved the species, and I don't think it was ever intended that he was to be an animal for food. As I recall it, they were taking the placenta with the embryonic life and maintaining life in it with a circulating medium the equivalent of blood. This was a test-tube creature produced outside of the womb of a human being. The growth of the embryo began in the human being or began with an impregnation of a human egg with a sperm that was not human, or vice versa, and then growth was developed outside the body." Myers senses the recollection that, as he says, "I participated in these things myself in a scientific capacity. I have been participating in strings of life. I've had lives of the most humiliating, debasing kind imaginable. But off and on throughout history there has been a scientific bent."

Though the hog, the upshot of these experiments, may have been accidental, Myers believes that "it was probably considered a great success just to produce the creature we produced. But the whole production might represent a span of 100 to 150 years. Whatever happened to the development of the hog in that period between the original plans and the success of it at the outcome I can't say. All I can say is I have strong feelings about certain animals having been developed from human flesh and thus relating to human flesh in such a way that we shouldn't use this flesh."

He is "under the impression" that seals were developed in this same way, and that it's possible the porpoise is another example. "This was the characteristic of animals that were developed from the human embryo—that they were far more intelligent than other animals." Myers, who is a vegetarian, declares that all of this is a compelling reason why people shouldn't eat hogmeat. "In terms of its taste and in terms of its molecular constituency, hogflesh is very near to human flesh. Then too, because the hog is a very intelligent animal, and also has strong feelings, the vibrations from the consciousness of the hog permeate its flesh. So the person who eats hogmeat is taking into his body cells which are going to bring influence that affects his thinking."

* * *

The truck rolled along a little faster than cruising speed over the barren narrow nightroad which had, by now, as it rose northward, replaced its flatness with hills and dips. In the rear, the hogs seemed quiet and more relaxed, some of them dozing on their feet and others savoring the belly tingle roller-coaster sensations of swoops and rolls and rises.

And up in the cab sat Hoover, with his feet propped against the dashboard, knees up, head lolled rearward and resting on the top edge of the backrest, as his glasses caught and reflected the glint from every flash of roadway light. He poised there briefly with mouth open, and then, compressing his vocal cords into the most pinched nasal bluegrass yowl that it's possible to squeeze from a human throat, sang forth: "*I lef' my ole hoooommme way back in the mount-tins . . . Muther was callllled to hea-vin that day.*" He swayed as he sang and rolled his eyes skyward with the soulfulness of a baying hound. "*They carried my muuuther up to the graveyard . . . Ever'thaing's loooonesum since she went away . . .*"

Then Moon chimed in on the chorus, and the two of them leaned their heads toward one another to howl:

> "*Muther's not deaaaad, she's only a-sleepin',*
> *Jest patiently waaaaaiiitin' for Jaee-zuss to come,*
> *Th' birds will be sing-in' while Mu-ther lies sleepin',*
> *They will sing oooer her as the grave sinks a-way.*"

After a tiny pause, Moon laughed and darted a foxy gleam sideways. "Man," he said, "*that's* what we oughta be doin'. Le's dump these damned hawgs an' try to make a go of it up in Nashville."

Hoover just snorted a little laugh in response and said nothing more; so Moon focused his attentions back on the road and, after a few minutes, began humming the tune quietly to himself in a self-contented tone. Meanwhile, Hoover's silence sank into a glum ponder. Moon's remark had jogged the hogs back into his head and, at the same time, filled him with the sudden stupid fear that if it were somehow left entirely up to Moon, Moon probably *would* dump the hogs—dump them on a railroad track or in a lake or anyplace. And the more he reflected on this, the more the old compassionate instincts of his childhood rose up inside him, for the first time in a long while, to the point where he began to catch fleeting tastes in his throat of the old tenderness he always used to feel toward pigs when he was little—and still *did* feel at times, though now it seemed so much harder to let anyone else know. And then, voicelessly, he declared to himself with a flicker of anguish, "There really *must* be some better way of doing this."

❋   ❋   ❋

The origin of the first relationship between men and hogs—like the earlier question of "Whence the hog?"—is another cryptic mystery, so far as rational, historical, scientific or otherwise scholarly sources of human wisdom are concerned. The question of domestication is simply another area in which the hog doesn't compute with the conventional machinery. The best assumption that scholiasts have made is that sometime within the Basal Neolithic period (5500–4500 B.C.) someone somewhere got hold of a hog and the two of them made a deal. Owing to the preponderance of pig evidence, it's been put forth that "China may have been the initial center of swineherding."[8] According to Darwin, the pig became domesticated around 5000 B.C. by the Chinese, who subsequently even went to the extreme of placing little model pigs carved from precious stone in the hands of their dead. Sometimes whole pigsties were buried inside Chinese tombs—the idea being that the departed shouldn't be denied this symbol of wealth and comfort in the next world.[9] All of which is made even more intriguing in light of the fact that China was supposed to have constituted, at one time, the eastern coast of Lemuria!

The question of when it was that domestication arrangements were first worked out between Man and Hog is important in that the nature of those arrangements formed the basis for all subsequent diplomatic contact between these two sovereign mammals. Essentially, domestication represents something of a treaty, a *détente*, a *rapprochement* between consenting intelligent creatures who enter into its special set of terms out of a spirit of shared respect and mutual self-interest. Yet it is these very terms which are now being called into serious question. Domestication, so states an 1819 volume, is "a tacit compact mutually binding betwixt man and the animals he domesticates, (and) implies a duty connected with an interest to both parties. Man furnishes to them food and protection, and enables them to pass a few years of comfortable existence: they repay him with their lives or their services."[10]

In theory, the first regular hog-human relationship came about when the most aggressive of the wild hogs, having initially gravitated toward man's storage sites for feeds and edible refuse, determined to set up general residence near human habitations, and then decided to stay and become companionable. "The timidity common to other animals was not shared by the hog, which had excellent defensive powers and feared neither man nor beast."[11]

Given all this, then, one might inquire into what motives or what specific needs on the part of each of the two creatures are satisfied by entering into a domesticational contract.

Insofar as the hog is concerned, it is suggested, to begin with, that the State of Nature is a condition of perpetual suffering and travail wherein there exist endless fears, threats, hardships of weather, and vagrant animals that just go around *biting* one another, all as part of a ruthless system called "survival of the fittest." The State of Nature, wrote Hobbes, is a situation of "continual warfare" in which the two cardinal virtues are "force and fraud." It is, in short, clearly unfit even for beasts, and hogs would do well to avoid it. Second, there exists among hogs the obvious need for food, shelter, creature comforts, sustenance for the aged and infirm, plus a congenial atmosphere in which to raise one's offspring. Third, there is always to be avoided, from the viewpoint of the hog (medicine not having been developed to a very great extent among them as of the time of their first contact with man), the dread and ravaging effects of disease. And finally, there lingers within the porcine soul the eternal quest for companionship and intellectual stimulation.

Man's rationale is more oblique. His first motive is diplomatic: Since the hog may easily be regarded as a numerical threat—what with humans and hogs being, respectively, the two most populous large mammals on earth—there thus exists a clear and present need for the two creatures to come to some kind of terms. Second, there's the problem of the hog's indisputable intelligence and the uneasiness this creates within man when he contemplates all that intellectual potential on the loose. Third, there dwells in mankind the unquenchable thirst for conquest, dominance and, whenever possible, physical exploitation of all other moving things. The fourth and most altruistic human motive—the motive most commonly found among Protestant missionaries, labor union executives, old New Dealers, graduate students and tumescent fat girls who write fantasy poems—springs from the desire to *liberate* the hog from the degenerate conditions of his life in the wilds, including the urge to shield him from those types of inter-animal relationships and activities that are viewed as "inhumane."

Having specified the respective needs which drew man and hog into common bond, the next point of inquiry has to do with the actual benefits accruing to the two contracting parties.

For his part, the hog receives protection from those other forest animals who may mean him ill. Further, he is provided shelter of sorts, usually including special places for sows to farrow (give birth) and nurse their young. And lastly, he gets a basic medical care plan that attends to his rudimentary health needs.

Man, for his part, gets the whole hog. With other animals, domestication is not nearly so demanding. For receiving approximately the same benefits as a hog is provided, the cow is called upon to donate milk; the sheep trades wool; the chicken lays eggs (which she doesn't necessarily want or need anyway); the horse supplies locomotion; and the dog offers up, in essence, a certain servile sycophancy which is translated as loyalty, love, trust and all those other good things.

But the hog gives his all—with no exceptions. And this being the case, it seems entirely proper to look again, a little more closely, at the nature of the hog's actual so-called benefits, this time viewing them in terms of the *spirit* with which man fulfills his contractual obligations.

The single hog, in good faith, enters into this joint arrangement motivated by his specific needs, and runs snout-first into these gruesome realities: His protection amounts to cramped confinement; shelter is usually minimal, non-insulated and oftentimes even nastier than a Mexican hospital; food is either rigorously monotonous or, on smaller farms, designated by terms of ultimate repulsiveness—i.e., "slop," "swill," etc. Swine families are separated, sexes are segregated and most boars are castrated. In precious few cases is the hog's natural quest for companionship satisfied; and in only one or two recorded instances has the intellectual stimulation provided by farmers—even the *potential* intellectual stimulation—been any more rewarding than what is normally available among hogs living entirely on their own.

Physically, the hog is removed from his ancestral wilds and weakened. He is deliberately drained of his natural litheness, fleetness of foot, pugnacity, muscle tone, bodily pride—of everything, in short, by virtue of which he endured under the "survival of the fittest" system. Fat is forced to replace muscle, and his natural organs of defense are rendered useless.

Out of some seeming man-bred malice toward his clean cut spirit and habits of fastidiousness, the hog is placed in environmental circumstances that would gag a goat, and is then compelled to conform his style of life to those very same squalid ghetto-status conditions that serve, thereby, to reinforce all the cherished human prejudices.

And yet . . . and yet . . . and yet . . . weakened, fattened, shorn of defenses, subjugated, humiliated, confined, abused, mud-flaked and ridiculed, he remains the quintessential hog: *sui generis, compos mentis.* He is persistently prideful, aloof, quietly defiant—and hence, in man's eyes, all the more exasperating. Man's most favored animals are those with a certain high-type, tuxedo-class social flair, creatures who may possibly generate moments of excitement, but no spiritual waves—like, say, race horses. For all their aesthetic flash, Thoroughbred race horses, as a class, have had all intelligence deliberately bred out of them to the point of their being utter brute beasts with mental powers barely sufficient to enable them to eat, drink and sleep of their own accord. Beyond that, if asked, they will run themselves literally to death without balking in the slightest—and, in all likelihood, without their even knowing what's taking place. Nevertheless, the standard terms of *their* domestic arrangements don't call for the termination of their earthly existences. Their docile stupidity (and hence the sense of smug dominance men feel in their presence) is the core of their charisma.

Hogs, however, know what's happening. And you *know* they know what's happening. And probably *they* know *you* know . . . etc. The upshot of which is that when a man finds himself in the company of a hog, those same warm familiar feelings of human-in-command somehow just don't flow. There isn't the slightest twinge of man-over-beast mastery here; rather the feeling called up instead is one of inexplicable malaise and vexation. Moreover, men often develop the bizarre sensation that their scrutiny is mutual: They get an uncomfortable *watched* feeling from the hog, made all the worse by a growing discomfiture and self-consciousness over whatever it may be that the creature is thinking. Or over the very idea that it thinks at all. (The average man is generally ill-at-ease around porpoises too.)

But more shattering for a man than an encounter with a single hog is his simultaneous haunting awareness of the sheer ubiquity, the omnipresence, of Hog in general. This uneasiness is compounded by the pervading sense of distrust men feel toward anything they can't control. In fact, in order to produce equivalent distress, dread and anguish of this sort it's not really even necessary to go out and confront hog-in-the-flesh. One can simply *think* hog and become every bit as unglued. For the very concept of hogosity disgorges a plethora of deep discombobulations in the tribal mind—wordless discomforts that seek to express (or perhaps resolve) themselves in the form of imagery and symbolic allusions. Not only has the hog long since ensconced himself into our culture on more levels of symbolism than any other animal but he also stands snout, jowl, head and shoulders above others in the outright quantity of powerful images (which is to say, passions in negotiable form) evoked in his name—images which are frequently at complete contradictory odds with one another. No one knows quite how to take him.

Consider the simple imagery surrounding the basic elemental hog. In this context, "Hog," to many people, means any obscenely rotund beast with a tropism for mud who trundles filthily along oinking, regardless of sex (there's been far too much emphasis placed upon sex recently). In contrast to this is the equally widespread vision of "Pig" as a pert, pink, bright-eyed, well-scrubbed, curly-tailed cuddly with impeccably pedicured trotters, who wrestles endlessly with such moral dilemmas as whether to go to market or to stay home, to have roast beef or have none.

Then there are the images relating to the hog's mentality, his relative wisdom and maturity. What somes to mind here is the animated film figure of Porky Pig: genial, red-jacketed, bow-tied and projecting "pig" as a stuttering fool, a buffoon, a blunderer, a Babbitt with hams. The diametric counterpart to him is the third of the Three Little Pigs (named "Practical" in the Disney version), who sagely built his house of bricks, then went on to dupe the wolf out of a load of turnips and finally caught his adversary in a caldron, cooked him, ate him and lived happily ever after.

In terms of alertness and pure intellectual energy, there are the pigs of political cartoonery cast in the role of utterly lethargic or loutish gluttons who are usually put forth to represent sleazy, bloated bureaucrats swilling up tax funds from the public trough (labeled "U. S. Treasury" or "The Taxpayer's Pocketbook" or some such). On the opposite pole there are the hogs of George Orwell's *Animal Farm*, who are depicted in light of their over-all intelligence, innovativeness and qualities of political leadership, if not domination.

In religion, hogs have sometimes been painted as creatures in league with—if not in the actual form of—the devil himself; then at other times they are creatures of great humbleness who, being simple and lowly, are considered pure of heart and therefore exemplary of the Christian ideal. Many ancient stained-glass windows in British churches show Jesus in the company of small hogs.

With respect to temperament, hogs, on the one hand, are the forbiddingly fierce-tusked beasts shown on medieval coats of arms and in pictures of Wild Boar hunts; while, on the other hand, they're the tenderly puffed-up cherubs of nursery room walls and fairy tales.

And finally, hogs have been imagined to have medicinal effects, both positive and negative, upon humans. Some ancients held hogs to be a cause of physical maladies in man, believing, for example, that pigs' milk brought on leprosy. On the opposite extreme, Egyptian physicians were advised by royal edict to prescribe potions containing the "blood, gall and liver" of hogs, all of which were considered to have powerful curative properties.[12]

Since hogs have been invested with more "meanings" and greater moral ambidexterity than anyone could ever possibly hope to keep up with, the real wonder is that people have yet still managed to consider them also as animals. The difficulty seems to be that beneath it all men view hogs as beings so disturbingly similar to themselves that the images of the two often overlap or become intertwined. Anthropomorphism abounds when it comes to hogs; they are a virtual conduit for people's feelings. For instance, it's hard to tell whether some of the hog expressions people use are intended to describe hogs or humans, or both interchangeably. To cite a few: "dirty" (or "greedy" or "fat" or "stupid") as a pig; "piggish" (or "hoggish" or "swinish"), meaning selfish or carnal; "pigheaded," referring to stubbornness; "making a pig" of oneself, meaning to overeat. To have been "brought up in a pigsty" means that someone is slovenly; to "live high on the hog" is to splurge in an overindulgent style of life; "to hog" is to hoard; "hog heaven" is supreme material bliss; and "pig," apart from its connection with the animal itself, can refer to (1) a printer or pressman, (2) a grossly undesirable girl, (3) a gluttonous man, (4) a foul, evil person, (5) a bar of iron or (6) a policeman. Now the main question that applies to each of these usages is whether the expression was coined primarily as an insult directed toward *humans* for acting some certain way . . . or toward *hogs* for the fact that undesirable people are drawn to parody and besmirch their patterns of behavior.

Underlying all this, though, is the fundamental fact that somehow the hog strikes sparks from some deep-recessed region of man's cortex which knows no language. The result is that men often perceive within themselves a half-suppressed kinship with the porcine race. At the very least, there exist today enough ritualized uses for the pig and a sufficient body of beliefs, superstitions and folk tales to suggest that an awful lot of people have an awful lot of faith that, in some way or other, something about the hog can and does affect the weal of human life.

People in Borneo still scrutinize pig livers for predictions about the future. Wealthy Moors in Morocco maintain Wild Boars in their stables to draw wandering evil spirits away from the horses and into themselves. The Tsembaga Maring tribesmen of central New Guinea consult their pig population to determine the timing of their rituals, the amount of acreage to cultivate and the frequency of warfare with other tribes. And in parts of Latvia, a pig tail is stuck into the ground when the first barley is planted to assure that the ears of barley will grow at least as long as the tail.

In the American South, eating hog jowl with black-eyed peas on New Year's Day is supposed to be a guarantee of good luck. Hogs, too, are considered weather prophets in many parts of the country. And here and there, a pious rural mother will assure her child that unless he or she does or doesn't do such-and-such a thing "the hogs will eat your feet." (Hogs being, by this interpretation, very wise, moral, all-seeing and, incidentally, fond of feet—just as we are of theirs, knuckles too.)

Other hog-based traditions carried over from antiquity include the stuffing and roasting of a Wild Boar's head at Christmastime in England, and the semi-ritualized slaughter, roasting ("in a kneeling position in the pan") and orgified eating of suckling pig.

Yet above and beyond any consideration of killing and eating a hog is the hope for all the spiritual gratification which can be gleaned from his vibrant *living* self. So imbued is Hog with portent and ceremonialized significance that the mere *proximity* of one (in, say, the adjoining room) creates, so some claim, a certain giddiness that can burgeon into a feeling of intense felicity. In short, hogs are a readily available source of rapturous mystical sensations—sensations which, for the avowed hogophile, can easily become intensified to the point of providing a cheap non-chemical "high."

As one crusty old rural New England hogophile put it: "Dogs look up to you, cats look down on you and pigs think you're their equal." The rural people of Ireland seem to recognize this equality and house pigs in their own cottages; "there he is emphatically 'the gintleman what pays the rint,' and is better treated often than the peasant's own children."[13] For pigs are beings of definite awareness, mood and sensitivity. In this vein, a very rare type of wild hog said to exist on a few islands in the Caribbean is reputed to be extremely retiring, shy and furtive in the presence of other creatures on account of his having developed, at some point in the obscure past, pubic hair. Natives allege that they have caught only the most fleeting sight of these hogs dashing shamefacedly from tree to tree, always seeking the proper covering and oftentimes carrying bits of bush or leafy vines trailing in their mouths in a vain effort to camouflage their hirsute genitalia.

Hogs appear possessed of a degree of honest emotional sensibility that clearly sets them apart from their domesticated cohorts of farm and field. A brief examination of the salient traits of these other beasts reveals the following serious drawbacks to their further consideration for a more elevated status in the wonderful world of animals:

MULES. Mules are excessively dour and ill-tempered. Few people report seeing a truly happy mule—an attitudinal problem generally thought to be the result of their profound and obvious sexual deficiencies. Mules are not only obstinate but tend to hold deep and often petty grudges, sometimes for years at a stretch, until finally venting their bile in some silly or violent way. Mules are presently on the decline, being phased out in favor of farm machinery.

COWS. A cow clumps along, chews grass, swishes its tail and *means* well, God knows, but insists on "mooing" in undifferentiated response to almost anything—"moo" being, on top of its other drawbacks, an absurd sound and virtually impossible to take seriously. (More about the cow later on.)

SHEEP. Sheep run in flocks and are, therefore, quite nearly helpless without shepherds assigned the task of "keeping watch o'er their flocks." Your female sheep, or ewe—being so very demure and docile and fleecy—is also something of a tease, coyly allowing herself to be regarded from time immemorial as an object for the passionate fantasies of untold hordes of farm hands.

GOATS. Goats seem on occasion to show promise and then, at other times, revert simply to being goats. At one point in 1962, there appeared to be a great potential breakthrough in the field of goats when a nanny on a farm in North Carolina began singing certain parts of *Aida* as well as specific roles in other operatic works by Verdi which, by some bizarre coincidence, had been the one-time hallmarks of a prominent Italian diva who passed away eight years before. Thinking the nanny to be a reincarnation of the late singer (and thereby opening the possibility of goats, as a species, being able to receive and express the artistic souls of yet other departed humans—or at least of Italians), a team of top-flight scientists, ventriloquists, astrologers and palm readers followed this singular goat for days on end, studying her and tape-recording her every aria until, at length, it was discovered that she had just memorized the whole thing off a set of LPs.[14]

So it comes back around to this: Through it all, it is the hog who remains the most richly dimensioned and enigmatic of creatures, even though he's among the most accessible to the average man. Perhaps humankind's subconscious fascination with the hog is rooted in the sensation that he (hog) seems far less like an animal, in the normal sense of the word, than are the other living things we are accustomed to dealing with in their undisputed role as animals. Perhaps the only real way of comprehending the hog in perspective is through the process of pitting him in comparison against every living creature and plant. In this way, you get to the hog not directly but by process of elimination, until what you eventually extract, in the final analysis, is pure distilled hogescence. But this could take years.

Animals that we regard wholeheartedly as *animals* may, indeed, be fascinating in many ways, but what it usually boils down to is a shallow fascination with their outlandish looks or totally non-human behavior—the sort of thing that people go to zoos to gape at. In this same spirit, citizens of Tasmania join platypus-watching clubs, which are said to have large followings.[15] But, again, the interest is in the *otherness* . . . not in the hopeful notion that the platypus has any points of commonality with man or anything buried deep inside to say. *Most* animals, in fact, amount to this—intriguing for the time being, yet for only so long as they remain colorfully unfamiliar, quaint or called by droll names that are fun to drop as long as they sound peculiar; to wit: Tokay gecko, tree frog, fox; bison, yak, wapiti, ox. Ibex, inchworm, squid, flamingo; dodo, tapir, dogfish, dingo. Bongo, bulbul, boa, bat; pocket gopher, water rat. Llama, puma, two-toed sloth; mandrill, gibbon, yucca moth. Emu, egret, 'lectric eel; drongo, dolphin, earless seal. White-browed gibbon, cynogale; mongoose, civet, finback whale. Sperm whale, humpback, Greenland, blue; pronghorn, boomslang, cockatoo. (Running through that list aloud about four more times will provide the average person far more exposure to any of these beasts than he ought to feel the need for ever again.)

Animal life to the purely urbanized man often just means cute things to squeeze. The hyper-citified mentality gravitates toward pastel images of little curly-locked moppets in big bonnets feeding geese from the steps of a thatch-roofed cottage, while two fat kittens peer out of an old boot. Urban animals are valued in terms of being flouncy, poofy, powdered, permanent-waved, beribboned, snuggly and small enough to fit into a standard bureau drawer. The currently favored pets of famous people, jet setters, celebrities and movie stars are: (1) lovely white Angora cats imported from somewhere (every one of them smug, and trained to exude the most utterly regal rottenness); and (2) toy poodles (perferably the neurasthenic, high-strung subspecies that spit up a lot—mostly on wall-to-wall carpets and expensive Afghans—and then refuse to cooperate with their psychotherapists); and finally (3) mynah birds (every single mynah bird ever born says the exact same thing). In addition, of course, it's said to be quite smart to have a lynx this year.

Now hogs can measure up to any of these city standards. Hogs can be neurotic, puny, dull, prissy, wrought with anxieties—all of those finer qualities of urban life. Yet it is by city people that they are continually made the brunt of the most malignant attitudinizing. As one prominent society lady, whose name you would know instantly, recently sneered: "It's no wonder to me that they call those completely disgusting filthy creatures 'pigs'!"

This anti-porcine outlook is, in point of fact, sufficiently strong to qualify its holder as a "mysosyst," an intense hater of swine. The sensitive man is hard put to understand a point of view of this sort, which even extends, in some cases, to a kind of libelous guilt by association as in the book *Hogs: Menace or Nuisance,* which featured, among other atrocities, several clearly retouched photographs of Hitler riding a hog. But even such hatred is a positive affirmation of hogritude; for in order that anything so personally remote could be so loathed it must first have managed to bore itself way down deep in the collective psyche.

At the opposite end of the spectrum are those stalwarts who eat, sleep, breathe and breed hog, particularly those folk out in the Midwest, America's most fertile hogland, who are avid believers in the old axiom, "Swine spell security." Pigs in these parts are frequently referred to as "mortgage lifters" owing to the fact that profits from Midwestern swine have traditionally paid off more farm mortgages than profits from any other class of livestock or agricultural crop.

The point is, no one can fail to feel *some*thing. Too much of the hog is obviously woven into the fabric of our society for any honest man to claim indifference. For, yea, the sole hog, with all the charm of a fallen archangel, *transcends* common everyday animality and speaks directly to forces within us: to hopes and hatreds, to violence and dreams, to passions aborted as well as fulfilled, to our spiraling moral intricacies, our trust in the absurd, our aspirations, to our fears spoken, and to those left unsaid. The passive pig may indeed be an ideal paradigm for modern man: man, like pig, who is a creature surrounded by filth and submerged in unavoidable dangers; who is a victim of circumstances created, in part, by himself; who is neither fully willing to do anything about his condition nor certain that if he *were* willing he would then be fully able —and on top of it all, who is, in some strange way, perhaps even capable of enjoying the whole thing. "We are sunk," wrote Schopenhauer, "in a sea of riddles and inscrutables, knowing and understanding neither what is around us nor ourselves."[16] All the really important things, even the obvious things, are mysterious and elusive. You can spend all day at Loch Ness and not see the monster.

Fact is, the hog and civilization are inseparable, for men and swine are eternally conjoined. We contrive, indeed, to attribute more of our own fouler human traits, failings and shameful self-perceptions to the hog than to any other animal—and then to hold him up for scorn and scoff and blame. Deep behind it all, then, what the hog may be condemned to represent is a dark, Dorian Gray repository for our utmost fears of the bestial side of ourselves.

Nevertheless, insofar as *he* is concerned, the flesh-and-blood hog has continually proved himself to be strong, moral and stoutly faithful to what he is: a creature hemmed in by his own existence and absolutely committed to it. Only a truly powerful life form could be able to absorb all this abuse and still maintain his integrity . . . and still be not less of a hog . . . and be not altered in his direction. Hence he stands as a constant, an indestructible. In contrast to this, unless humankind manages at once to turn around some of its *own* basic propositions we soon may be too far gone to rescue. But the hog will endure—and, by some chance, may perhaps even prevail!

He is thusly a study in values. Or a study in something. Perhaps it's the sensation of sublimity, since the sublime is that which awakens inside us the idea of the infinite, and arouses feelings of awe. Still, for now, for us, the hog exists as a looming blur. And hogness remains a vast vacuum into which we project ourselves . . . and then peer hard inside to try to glimpse what fearsome, infuriate thing has been formed.

*　　*　　*

Tirewhine on asphalt. The blacktop road had dimmed into a windless, restive quiet as the truck droned its way forward along a corridor of somber trees. In the truckbed, the hogs stood all awake and altogether silent . . . with their inertia generating a certain listless ominousness which moved forth slowly in the air like a baleful amoeba. In time, it crept up front into the cab. (And Moon felt it first and cleared his throat and said—mostly just to reassure himself that his voice still worked— "Lordgod, Hoover, I *know* there's gotta be some better way of gettin' this whole business done.")

Outside, a new wind blew out of the black and down the long canyon of passing trees. Spirit things lurked behind those trees: spectres, ghosts, bogeys, banshees and vague invisible forms that harry the land at night. Ogres, ogresses, vagrant apparitions, daemonic shapes . . .

In the lightless air of the truckbed all the labyrinthian mysteries of hogritude—ever so softly quivering at first—began to move. (And up front, Hoover nodded and said in a muffled tone, "Lotta animals know a lot more than they're tellin'.") And soon the truckbed—which had fallen grimly stilled some distance back—with a little tremor, grew soundlessly alive: now became a grunting, undulating agglomeration of hogflesh softly rippling, as across the surface slithered stirrings from something crypto-porcine, primal. (And Moon, with eyes and hands on the wheel and face aimed straight frontward, shook his head for a full half mile and —making each word carefully distinct, but not too loud—declared, "Hoover, I think we're in the presence of somethin' or 'nother a whole lot larger'n ourselves.")

Far down beneath the earth huge forms moved silently and arrayed themselves. Trucktires whined on. And from the bed there rose a gently muted nuzzle . . . rustle. A reaching lurch. And then an ever-so-faint searching sound (or indistinct *feeling* as if of sound) abruptly fading away, sailing free, growing distant—a fading . . . stretched and gliding . . . sound-sound-sound, like unto the soft slowdeep flap from gigantic wings, or the spirit of 45 million years' worth of hogs rising into the heavens, renewed.

II

Hogs Historical

Shambled rural homesites slid quickly past the trucksides. Roadside signs rose screaming out of the night, then slunk back to where they lived. Trees streaked by: trees and forests across whose floors moved hushed darting animals.

If left to their own devices, the hogs back in the bed of the truck would gladly and quite naturally revert to the freedom of pure forest life. The primeval heritage of hogdom is rooted in the earth's dark wilds and woodlands, such that at their innermost core all domestic hogs are at least potential forest beasts. They originally inhabited forests, and many of them still do—like the Wild Boars who typically dwell deep away in woods and come out only in the evenings to grub up anything edible in the vicinity.

Generally, your modern Wild Boar is a sociable creature who thrives away peacefully in herds, maintains close family ties, unhesitatingly defends home & hearth, washes regularly, keeps sensible working hours and otherwise does all the same respectable things the way they were done before man arrived. Basically, all hogs were at one time just as wild hogs are today—lean, clean and happy. Now, though, what with the influence of "modern man" and "scientific swine management," hogs, in their domestic form, have developed into creatures commonly afflicted with gastric ulcers.

But man's dominion has not yet spread so far as to humiliate and demean that entire branch of animality that has historically been included within the great world of Swine. And indeed, it *is* a great world, whose domestic membership alone stands at somewhere over 520 million —with virtually countless others functioning away unfenced far out in the bush and in places where people rarely tarry.

The multitudinous wild, unfettered members of the swine family currently hold forth on every continent except Australia, as well as on most of the temperate islands in both hemispheres; and since they persist in conducting themselves in pretty much the same style now as in the Beginning, those wild ones that abound in our midsts deserve special attention not only as contemporary legitimate creatures in their own right but also as possible clues in answer to the swelling question: Wherever did the hog come from—and where is he going?

The principal variety of untamed pig—the species of hog-in-the-raw so popular among hunters, heraldic artists and certain jaded gourmets— is the European Wild Boar (*Sus scrofa*), who ranged all over the place, from Europe to Central Asia, from the Baltic to North Africa, as well as in some parts of North America. The Asiatic Wild Boar (*Sus scrofa cristatus*) is a subspecies that operates out of Ceylon, India, Siam, China and various spots within that general region. All the breeds of domestic hogs were originally descended from *Sus scrofa* and are fully able, even now, to have successful sexual relations with Wild Boars—and do, whenever the opportunity presents itself.

Far from being physically ponderous, the Wild Boar (which is sometimes called the Prussian Wild Boar) is a sleek, stout, well-proportioned beast built for running, fighting and executing all manner of vigorous porcine feats. He is usually about three feet tall at the shoulder, weighs 350 pounds and wears a dark brown coat with no hat. He has small bright eyes, ears that are large and always pricked and a tail that hangs down straight and dignified, never coiled. His snout is, of course, the

European Wild Boar

standard one with which all hogs are outfitted; but rising up on each side of that wonderful snout, and curving backward, are his two lower canine teeth, which constitute formidable tusks eight or nine inches long. Because of these tusks, or tushes (plus his native litheness, swiftness of foot and innate sense of military stratagem), he has always been considered among the world's most dangerous game—being able, among other things, to bite more savagely than a tiger, lion or bear. "Only the Killer Whale of all the earth's mammals can inflict a worse bite."[1]

But for all this capacity for ferocity, he is an instinctively gregarious and goodhearted fellow. Most Wild Boars tend to live quiet day-to-day existences in herds which consist largely of the females and young under the guidance of an old sow. Only the older male will sometimes live solitarily. Yet come nightfall, he and all the rest will steal softly in the moonlight from their woodsy private lairs to the nearest patch of cultivated landscape to gouge the ground, root, eat and generally browse happily away till the crack of dawn. Wild Boars, like domestic hogs (and men), are mostly omnivorous, but are said to be "particularly fond of beetroot, potatoes and Jerusalem artichokes. Above ground they gather beechnuts, chestnuts and any kind of fallen fruit, but they also feed on insects and their larvae, earthworms, reptiles, birds and their eggs, and small mammals. They appear to be immune to snakebites."[2] Most usually, just before or after eating, they will make their way to some favorite lakesite or swampy pond for a pleasant wallow—this being necessary not only for peace of mind and to relieve inner tensions but to rid themselves of parasites. Wild Boars are scrupulously careful, however, not to despoil their wallowing spots. They actually prefer clean water as opposed to mud. They also never contract ulcers. This creature, then, *Sus scrofa*, can be considered more or less as your most basic wild pig with all the preferences and attitudes and styles of living that domestic hogs would affect if they were given similar free rein.

The world's largest hog is probably the Bornean Pig (*Sus barbatus*), who is six feet long, nose to tail, with an enormous head, a crest on the crown and a bristly moustache that camouflages his tushes. Smallest is the Pygmy Hog (*Sus salvanius*) of Nepal and the Himalayas, who stands about a foot tall and whose piglets at birth are each not much larger than a package of cigarettes. The most predaceous, rarest and least known about is the giant Forest Hog (*Hylochoerus meinertzhageni*), who lurks and looms among the Ituri forests and throughout much of the rest of equatorial Africa. No white hunter is known to have killed one of these; and in fact, the giant Forest Hog was only a subject of dark rumors before one of them was finally nabbed by natives in 1904. The giant Forest

Wart Hog

Hog is jet black with a long body almost the size of a pony, massive head and warty growths below the eyes, making him, in sum, just a half a shade less ugly than the Wart Hog. It must be acknowledged that the African Wart Hog (*Phacochoerus aethiopicus*) is the ultimate in Ugly, with a misshapen head, wart-like lumps on his cheeks and grotesque tusks. And as if he weren't repugnant enough already, the Wart Hog calls further attention to himself by running along with his tail up stiffly erect, and then, when confronted, will frequently fly into total hysteria, trampling frantically over other (and equally ugly) Wart Hogs in his escape.

The Red River Hog (*Potamochoerus porcus*)—typical of a wide variety referred to as "African Bush Pigs"—wears a coat of reddish bristles with white muttonchop-type whiskers and dwells in the deep marshy forests of southern Africa and Madagascar. The next-closest-by species makes its home in eastern India. This is the largest-tusked hog, the Babirussa (*Babyrussa babyrussa*), or "stag hog," whose lower canine

Babirussa

teeth grow straight up and then back almost in a circle, looking altogether so unearthly that the natives maintain that these pigs hang themselves by their tusks on tree limbs at night.

The principal New World counterpart to all these European, Asian and African pigs is the Peccary (*Tayassuidae*), found all the way from the wilds of the southeastern United States to Argentina. He is fairly small, thin-legged and, according to most reputable texts, is not really regarded as a "true pig" anyway (by the Rules Committee, or whoever decides on these things). There are, just so you can keep them straight, two types of peccaries: Collared Peccaries, or *Javelinas*, who are blackish-brown with white collars; and White-lipped Peccaries, who're more reddish-colored and are equipped, just as you would suspect, with white lips. Peccaries are not much larger than Pygmy Hogs and therefore find it advantageous to travel in bands of fifty or more so as to be able to gang up on other animals, or men. Perhaps it is on account of this very unsporting, un-hoglike method of operation that they are not considered "true pigs." In any event, they contribute nothing to our understanding and will not be dealt with further.

Another creature that's not a pig is a hippopotamus. Of all other animals, however, hippopotamuses are the closest kin to hogs, being the only other even-toed, non-cud-chewing member of the order *Artiodactyla*. Then too, hippo meat is supposed to taste something like Wild Boar. Other than that, the most salient feature the Great African Hippopotamus shares with the hog has historically been the degree of misunderstanding and sneering abuse heaped upon it by civilized man. Back in the days of traveling circuses the hippo was a popularly featured item of sideshow disgust, usually billed as the "Blood-Sweating Behe-

Collared Peccary

moth." This designation was derived from the fact that the skin pores of the hippopotamus normally exude a pinkish insulating fluid known as "blood sweat"—which consequently meant that an actively perspiring hippo seemed to display the chief element that has always been sure to draw crowds. Other than that, hippos didn't do much and never really caught on with the population in general.

Now the question that quite naturally arises at this point is: What has all of this got to do with hogs historical? It is this: All of the creatures discussed thus far are hogs in their original, natural and most-nearly-primitive state, untouched by human hands and mostly unaffected by the ravaging course of human history. (In all fairness it should be acknowledged that there is a theory holding that hogs are not actually of this earth at all but are from UFOs, having been implanted here by extraterrestrial beings as some vague sort of experimental project to see how we would be able to adapt to and share the earth with a creature so much like ourselves, and not merely exploit him. There is some merit to this, at least in terms of the fact that it can't be absolutely disproved, what with there being no compelling evidence as to where *else* the hog might have come from; but for the moment, at least, we are going to proceed along more accepted academic and historical lines.)

In tracing out the riddle of the hog across the ages, our perspective most properly begins with the Eocene ("dawn of the recent") Epoch, which opened, to mixed reviews, about 45 million years ago. It was followed by the Oligocene ("little recent"), the Miocene ("less recent") and Pliocene ("more recent"). Then came the Pleistocene (or "most recent," during which time-span man first occurred), the Holocene ("wholly recent") and Just-the-Other-Day.

It was during the Eocene Epoch that there came forth, from somewhere, gigantic pig-like beasts called entelodonts, who—or so it appears from evidence that has since been discovered in Europe, Asia and parts of the western United States—ran madly all over the place making fossils of themselves. These animals, which stood over six feet high at the shoulder, became extinct after a while, probably on account of the fact that, being so big, they ate up everything around them and then fell to quarreling among one another—as is so very often the case with men and nations. If we accept these as proto-hogs, then hogs have been around for at least 45 million years, which clearly implies that hogs were hogs long before men were men.

But appearing on earth right on the heels of the entelodont was the Tusked Hog (*Palaeochoerus*), who sedulously thrived away up until the beginning of the Miocene Epoch, 19 million years ago. By that time he had evolved sufficiently so as to be reclassified as *Hyotherium*, whereupon he promptly continued his self-modifications and evolvements at such a breakneck pace as to qualify himself eventually and at last as the official precursor of *Sus scrofa*.

Then one morning some millions of years later, all the world's creatures crawled drowsily out of bed, just as they had always done, and suddenly now found themselves in the Pleistocene Epoch. Naturally there were some pretty surprised expressions on a lot of faces, particularly a bit later on in the epoch when Man made his initial appearance. Actually, he was a "hominid," or man-like creature, but still he fit the description more closely than anything else around. This was Java "man," who, after a time, relinquished the title to Peking "man," who, in turn, passed the honor along to Heidelberg man, then to Ngandong man, to Neanderthal man, to Cro-Magnon, and so on and on up through history—past Charlemagne, Prince Metternich, Oscar Wilde—progressing along into contemporary times, most recently to one Walter G. Lampley of Wayne, Nebraska, where it now rests. But this is getting ahead of the story.

The Pleistocene Epoch, which commenced about a million years ago, marked the advent of a man who, not very long after he arrived, entered into the Stone Age. Everyone, of course, is by now perfectly well aware of what man was like back then. He was awful. He considered himself to be doing good just to stand upright. He spent most of his time in dim caves quaking in shameless fear of wild beasts and even wilder gods. Yet no sooner did he get himself somewhat firmly established on the earth than he struck up an acquaintanceship with *Sus scrofa*. Evidence found in Europe dating back 600,000 years shows that Heidelberg man ". . . shared the fields and forests along with the Wild Boar, the broad-faced moose and the straight-toothed ancient elephant."[3] Later on, the hog was associated with the Neanderthal Man in Germany and after that, in France, with the Cro-Magnon.

The Cro-Magnon was notable as a hunter but more especially as a man of some artistic pretension who was able to paint things on the ceilings of caves and certain subterranean and evidently sacred places where no one else would normally venture to look. About thirty thousand years ago, at Altamira in northern Spain, some Cro-Magnon artists depicted several Wild Boars on a cavern roof, plus one particularly ominous boar in fierce mad gallop across a wall. This is the earliest known picture of a hog. It was right *after* this particular endeavor, however, that the Cro-Magnons were conquered by a Neolithic people who had just recently stumbled into Europe from the East. (Actually, it was about twenty thousand years after; but in these vast and hazy reaches of prehistoric time thousands of years here or there aren't really that important—it's the *feeling* of the whole thing that counts. Just think of it all as a long, long time ago.) So here came Neolithic man, some seven, ten or so thousand years ago, accompanied into Europe by herds of hogs that he had gradually domesticated during the course of his wanderings—making *him*, therefore, mankind's first breeder of swine.[4] After this, Neolithic man went on to make pottery, textiles and polished tools and to invent the bow and arrow and the wheel. But these feats tend to pale by comparison.

According to Darwin, the pig became domesticated in China about 5000 B.C. It was not, however, until 3468 B.C.—when Emperor Fo-Hi issued an imperial decree ordering that swine be bred and raised—that there appeared the first official written evidence of Chinese hog husbandry. China became then (and remains today) the world's top-ranking hog-raising country. The ancient Chinese subsequently succumbed to such a swine fondness that, as mentioned earlier, they sometimes included whole pigsties in the tombs of their well-to-do citizens. Indeed, in China, as elsewhere, the hogs themselves rapidly slipped into the curious dual role of being creatures bred both as ordinary livestock and as semi-holy objects of awe to be served up in sacrifice on periodic occasions of profound religious ceremony or significant commemoration.

Beyond the boundaries of China, archaeologists have excavated man-made hog memorabilia from all over the ancient civilized world, the earliest being one at Anau in Russian Turkestan, dated about 6500 B.C. In Egypt, little clay models of pigs were found in graves dating back to some obscure period before 3400 B.C. The earliest mention of swine in Egyptian literature occurred about 2900 B.C. The political recognition of hogs was made even more official in 1980 B.C. when King Sesostris I created the office of "Overseer of the Swine." And their spiritual qualities were acknowledged a bit later on, around 1300 B.C., when Egyptian temple priests began pig-breeding in earnest. (More about this in the chapter on hoglore.) Pictures from about that same period, both in Egypt and Greece, show that hogs were commonly called upon after floods and rains to tread grain seeds into the earth. This was because their pointed hoofs made holes of just the right depth to assure germination.

Greek artists also painted pig pictures on vases and other such items to illustrate the various local stories and legends of the time that frequently involved hogs. Hog-related historical legends became popular among the Romans too. Aeneas, the Trojan hero and legendary founder of Rome, knew that he had reached Italy when, as prophesied, he saw a great white sow with thirty young by the River Tiber. In Book viii of *The Aeneid*, Virgil tells of how a deity advised Aeneas to recognize the site whereupon to build the city that later founded the colony that became Rome:

> And that this nightly vision may not seem
> Th' effect of fancy, or an idle dream,
> A sow beneath an oak shall lie along,
> All white herself, and white her thirty young.
> When thirty rolling years have run their race,
> Thy son Ascanius, on this empty space,
> Shall build a royal town, of lasting fame,
> Which from this omen shall receive the name.[5]

The "Great White Sow" became an important symbolic feature of Roman tradition. From that time on, pigs, both wild and domestic, appeared regularly in Roman life and art and even on coins.

What with hogs being so socially acceptable, if not officially sanctioned, the Roman Twentieth Legion adopted a Wild Boar as its emblem and went on to spread models and drawings of it all over Britain. The Britons, however, had been breeding *Sus scrofa* since 1500 B.C. Celtic legendry of that time recorded that "there have come to the south some beasts, such as were never known in the land before, and their flesh is better than the flesh of oxen."[6]

Historically speaking, the shift from the nomadic existence of a hunter to the more settled role of a planter of food and breeder of livestock was the most important step in the cultural ascendance of man. Agriculture was the basis for civilization, and the hog was a predominant part of the basis for stable village agriculture. Whereas the horse, for instance, was initially domesticated and employed for purposes of warfare, the hog has always had only the most benign domestical relationship with humanity, to the extent that his mere presence has always connoted peace, stability, happy homelife and prosperous fecundity. For the most part, the hogs that ancient man domesticated were slightly modified Wild Boars, who, owing to their innate integrity, managed to retain most of their own admirable characteristics as well as the enduring respect of their keepers. This respect was shown in a plethora of early pig portraiture and depictions of pigs in precious stone. There are pigs in jade from China, ebony pig figurines from Egypt, bronze boars from Roman Britain, pigs in ivory and gold, pigs woven into tapestry, pigs in the form of puppets made of buffalo hide from Java, pigs in terra cotta, pottery, glass and porcelain. In fact, "porcelain" comes from the Italian word *porcella*, meaning a young sow.[7]

But the rigors—or actually the lack of rigors—of later domestication played hell with the proud hog. One reason he had taken so well to domestic arrangements to begin with was that he was intelligent enough to be able to adjust himself to living amicably alongside a different type of animal. Yet as time went on this required that he adapt himself more and more to constricted situations wherein his old warrior qualities of litheness, speed, strength and resourcefulness were unable to be exercised. As steadily greater amounts of land were put under cultivation and made off-limits to him, and as he came to be confined within ever smaller enclosures, the hog's powers of endurance and physical defense became less developed, his muscle tone progressively succumbed to meaty bulk and fat, and, in time, even his nobly aggressive disposition changed—as also, eventually, did most people's conception of him. Thus it is that domestication—at least the way it is currently comprehended and devised—becomes a one sure path toward degradation, and is the ultimate snare for all abundant animal spirits, and is the kiss of Judas.

With the pig, in the context of human civilization, being regarded increasingly as an article of docile livestock rather than as an agile hunter and occasionally fearsome adversary, the very word itself came to be used more frequently with all manner of derogatory overtones, most of which were simply due to people's ignorance of the pig's basic character and personal preferences. There are, in fact, more hog-related expressions than expressions involving any other animal—at least in the English language—and most of them have to do with linking men to hogs. And sometimes they are matters of great seriousness. Erskine May's handbook of parliamentary procedure, which is the official reference for the British Parliament, lists the names that members are specifically proscribed from calling one another. They include: guttersnipe, cheeky young pup, rat, stool pigeon and swine. Then too, there was the historical fact of World War II's diesel-powered submarines being nicknamed "pigboats." The explanation, according to one old submariner, "would be instantly apparent to anyone who ever stood downwind of one that opened its hatches after it had been submerged a few days."

The roots of the word "pig" are lost in antiquity, but the sound itself is actually euphonic and appealing, especially in its earliest forms. In Low German it was *Bigge*, in Middle Dutch, *Vigghe*. Then in Old English it was *picga*, in Middle English, *pigge*, and from about the sixteenth century onward, pig.[8] Now, the origin of "hog"—which is used more or less interchangeably both in reference to the animal and as the basis for many an unkind expression—hasn't yet gotten lost, but there's always the lurking possibility that it might. "Hog" is said to be derived from a certain Hebrew word that meant to *encompass* or *surround*, "suggested by the round figure in his fat and most natural state." Other hog historians, however, "are more inclined to refer the Hebrew noun to the Arabic sense of the verb, *viz., to have narrow eyes.*"[9] Take your pick.

In terms of name, background, development and practically everything else, pigs were strictly of the Old World. But they were among the very first Europeans to set hoof upon the New. At the insistence of Queen Isabella, Columbus ferried eight porkers across the Atlantic on his second voyage in 1493. From these were descended all the hogs that soon populated the islands of the Spanish West Indies.[10]

The next load of hogs came with Cortez, who brought them to the American mainland about 1521. Since it seemed practical at that time for invaders to come equipped with their own meat supply on-the-hoof, Cortez's overland expeditions were usually brought up in the rear by an ever present drove of swine. During the Spanish colonization movement that followed, hog-raising was officially encouraged in all the settlements.

When Hernando De Soto landed in Florida in 1539, he brought about a dozen hogs he'd picked up in the West Indies. Over the following three years, De Soto's band trooped all across the territory of what is now Georgia, North and South Carolina, Tennessee, Alabama, Mississippi and Louisiana. The main thing all this accomplished was that it exposed hogs to the Indians, or vice versa—which, in a few instances, led to the Indians' becoming such bedazzled addicts of hogmeat that they resorted to hog rustling and even to minor warfare. Here and there, however, as the expedition moved on, Indians would legitimately come upon hogs that had escaped and would proceed, thereupon, to become hog breeders themselves and resign themselves to more peaceful ways.

The British colonists who settled in Jamestown in 1607 brought a handful of hogs which proliferated away, as is their wont, to the point where soon the place was practically overrun. And here, too, the local Indians took such a liking to the creatures that the colonists felt it necessary to confine their herds to Hog Island, where they were then free to continue their reproducings at an even greater pace. Eventually, feeding these growing battalions of hogs became such a problem that they were set loose in the woods on the mainland to forage for themselves. By 1627, a local livestock census was able to determine the precise number of cattle, horses and so on, but "the swine population was innumerable."[11]

To the north, a detachment of hogs landed with the Puritans as part of the Massachusetts Bay Colony. And naturally they started procreating madly. And naturally this got into a thing with the Indians. After a while, laws were passed requiring the Englishmen to mark the ears of all hogs to distinguish their own from those of the Indians. In very short order, hogs became an excuse for taking over Indian territory. The tribes in New England depended on agriculture for their subsistence, and whenever it happened that colonists' hogs were caught rooting up a crop the Indians would quite naturally kill them. King Philip, chief of the Wampanoags, described how these hog killings were used as a device for occupying more and more Indian land: "Sometimes the swine of the English would come into the cornfields of my people, as they did not make fences like the English. I must then be seized and confined until I sold another tract of my country for satisfaction of all damages and costs. Thus tract after tract is gone."[12]

Around 1690, William Penn's colony of Quakers—who had started out a few years earlier employing all the current principles of sound swine management—had become so hog-rich that they were able to export hog products as well as live hogs to other areas. Penn opened livestock markets in various parts of Pennsylvania and even began holding annual hog fairs.

But the real advance of American hogdom took place during the first half of the nineteenth century, after the federal government had nabbed sufficient western territory from Indians to make new land available for settlers. In packing for the trip, most of these homesteaders quite naturally decided to take along a set of hogs, who as it turned out, were particularly well suited for rigorous peregrinations across raw terrain. For one thing, hogs thrived and proliferated better than other animals and displayed whole arrays of abilities and practical applications. Many more than, say, cows—whose chief talent, apart from the giving of milk, has historically been to serve as a unit of time measurement, as in ". . . from now till the cows come home." (But more about cows later on.)

The hog who went west was not only hearty and pleasant to be around but perfectly able to defend himself against coyotes, foxes, wildcats, snakes, wolves and even a vagrant bear or two. As an English observer described these old frontier hogs: "They are long in the leg, narrow on the back, short in the body, flat on the sides, with a long snout, very rough in their hair . . . You may as well think of stopping a crow as these hogs. They will go to a distance from a fence, take a run and leap through the rails three or four feet from the ground, turning themselves sideways. These hogs suffer hardships as no other animal could endure."[13]

This creature who trekked into the unprobed boondocks with settlers and frontiersmen in pursuit of the nation's (and his own) "manifest destiny" was basically a "woods hog," so called because he was able to feed himself by foraging in any forests that happened to be convenient. He was also affectionately, and at other times abusively, assigned a number of other tags, many of which have hung on through the years. He was known as a piney woods rooter . . . a stump rooter . . . wood wanderer . . . landpike . . . prairie racer . . . mountain liver . . . acorn gatherer . . . bristle bearer . . . alligator . . . hazelnut splitter . . . wound maker . . . and razorback. Counting the roughhewn hogs of this type, along with the more genteel domesticates back east, the U.S. hog population by the mid nineteenth century was upward of 10 million, or roughly half the U.S. human population as of that time. About that time, too, hogs unwillingly became a means for another kind of exploitation and political chicanery. A European commentator on the American scene during that period observed: ". . . a barbecued hog in the woods, and plenty of whiskey, will buy birthrights and secure elections in America."[14]

Pigs, bless them, were everywhere. The early 1800s were, in fact, something of a golden age for the porker in America. Hogs were even allowed to parade freely on Broadway. At one time, before they thought better of it, New York authorities allowed hogs to roam and root for themselves on the city streets. From the porcine point of view, the chief occupational hazard in this was that of being pursued and perhaps killed and carried off by any interested citizens—an activity that was much applauded and avidly egged on by the local press. One reporter described a frantic chase "down Broadway, the thoroughfare of fashion, taste and beauty," after it happened that a "good-sized pig was observed to be quietly perusing the remains of some offals, which not infrequently adorn our principal promenades."[15]

Apart from his offering himself as a source for recreation and deep companionship throughout the course of American history, the hog has incidentally played a vital social role as a provider. Salt pork was the one essential form of meat carried not only by explorers and frontiersmen but also by U.S. soldiers. In the winter of 1776–77, salt pork from southern New Jersey was slipped into Pennsylvania by night past British sentries to feed Washington's troops at Valley Forge.

On one occasion during the War of 1812, a New York pork packer nicknamed "Uncle Sam" Wilson had shipped a boatload of several hundred barrels of pork intended for American soldiers. Each barrel was stamped "U.S.," which prompted someone, seeing the boat at the dock, to ask the watchman the meaning of the letters. The watchman answered, "It must mean Uncle Sam . . . Uncle Sam Wilson—he's feeding the Army!" In practically no time the phrase about Uncle Sam feeding the Army became so well known that the tag "Uncle Sam," plus the newspaper cartoonists' image of him, came to personify the government itself. (As can be seen, none of this would have come about were it not for hogs.)

Pork has actually been a basic ingredient in the battlefield provisions (such as in C rations, K rations and such) of American soldiers in every war since the Revolution.

I frankly don't know if this holds true for the soldiers of other countries, but I strongly suspect that it might be so for the British. Traditionally, the British are just fools for hogmeat. (One notes in the Guinness Book of World Records [9th Edition] that Rodney Harrison of Shropshire, England, set a record in 1969 by wolfing down twenty-five two-ounce sausages in four minutes thirty-seven seconds. Furthermore, the world's longest sausage was made in Scunthorpe, Lincolnshire, England, in 1966 and measured 3,124 feet.) A particularly memorable function played by the hog—or his essence—in British history revolved around the occasion when the Moslem soldiers in India refused to use certain new-type English-made rifle cartridges because they were rumored to be greased with pigfat (though in fact they weren't). This touched off the 1857 Sepoy Mutiny, which ended the 250-year political domination of the British East India Company.

❅   ❅   ❅

Past a certain point fairly early in the ascent of intelligent life on earth, hog and human history become so intricately interwoven that the student who is devoted to an honest, holistic understanding of things-in-general necessarily comes to accept the subtly suppressed truth: that hog and human development have advanced hand-in-hoof, so to speak, in a symbiotic sort of equilibrium, and that were it not for the porcine race the course of man's own history would have been considerably altered at several points along the line, more than likely for the worse. The hog's significance as a stand-in for man, symbolically and otherwise, has manifested itself throughout the past on all sorts of levels and dimensions.

Early European explorers who ventured to the islands of the southwestern Pacific continually came upon—and occasionally fell prey to—deeply primitive tribal cultures in which cannibalism had always been an accepted feature of life. In an effort to refashion local tastes and bring the natives into line with Christian ethics, British missionaries, operating among the islands of Melanesia in the 1830s, introduced the population to European domestic hogs. Here was a meaty creature the natives could raise and eat and feel good about. And they did. The tribespeople almost immediately affected a great and abiding fondness both for the animals themselves and for their palatability. Hogs became a form of currency and a measure of wealth and an indispensable part of their religious rituals. And the tribesmen became the most zealously devout hog farmers in the world—but they also continued being cannibals. Only from then on, they referred to human meat as "long pig." This was on account of its profound indistinguishability in taste and texture (and vitamins) from the flesh of the shorter pigs they'd so recently come to love.

While it's hard under normal circumstances to uproot anyone who is willing to partake of and comment on the question of human versus hog-meat, I did at least discover a man who had done so at one time and who was not loath to talk about it. The writer Anthony Burgess recounted to me a personal experience that took place when he was serving in an official capacity with the British government in New Guinea shortly after World War II.

"On this one particular occasion," he said, "there was held a large and very important ceremonial feast which I attended and took part in. Naturally I made it a point to take a sampling of whatever foodstuffs were offered to me. I remember being served a portion of meat which I ate and found to be quite good, very much like a fine, delicately sweet pork, which is what I thought it was. When I was told that it was the meat from enemy warriors who had been killed in a recent tribal skirmish I almost immediately voided, threw up, that which I had just eaten. But I do remember what it was like, and I might just as easily have gone on thinking it was the meat from a hog."

Burgess went on to recall that the day after this incident he was shown some large enclosed sheds in one of the villages containing "bodies of slain enemies hanging by their ankles from crossbeams overhead, with their drippings falling on beds of vegetables below." These sheds, he learned, were regularly used for curing new meat, with the bodies left just to hang there like so many hams or sides of pork, developing savor.

The amazing interchangeability of hog and human roles in symbolism and legendry and ritual is part of the whole vast trove of hoglore which is dealt with later on. But for now, for purposes of appreciating porcine history, it becomes easier to comprehend and relate to the plight of the hog throughout the ages if one accepts the actuality of the simple physical fact that *we are pork!*

The proper focus of this story, however, remains the live, vibrant, prideful, up-and-grunting hog and his place in the annals of time and what all man has made of him and done to him, and vice versa. In large measure, man today, at the crest of this final downhill slide of the twentieth century, behaves as he does because of all the legions of influential acquaintances he has made upon this planet ever since the Dawn of Time. And one of the oldest, most loyal and most consequential of these has been the common hog.

The mere concept of "Hog" allows, even compels, men's minds to dig deep into those private fears and glees buried far, far back in the tribal subsconscious, back to our earliest and therefore most shared shelves. The hog shows us things. He shimmers with implications. For instance, during the 1946 test of an atomic bomb on Bikini Atoll in the Marshall Islands, a fifty-pound shoat designated as "Pig 311" fell, or jumped, from a ship in the test area at the moment of the blast and managed to swim through radioactive waters until she was rescued. Pig 311, whose case became famous, lived out her later years as a normal six-hundred-pound sow in Washington's Zoological Park—except for the fact that she was forever inexplicably sterile. All of which explains why so many people since that time have had disquieting dreams at night.

Yet for all the hog has done and has been over the aeons, for all those hogs, both friendly and foe-ly, with which humanity has shared direct contact, contemporary civilized urbanized man still resists coming to terms with basic hogritude. The mandarins of the cultural *status quo* have simply never come up with a formula, technological or otherwise, that fits. For the most part, the hog occupies the shaky ground of a vaguely sacred cow: untouchable, but for reasons that somehow no one has ever made quite clear. He remains a subject that summons up chuckles and rue and untold rumination. Though he has etched his cloven hoofprints upon the face of Old Mother Earth, starting forth from the darkest and most backward abysms of time, and has been a legitimate historical entity in his own right, the hog is not to be regarded as a phenomenon who has had his day, a beast merely of the murky past. Nor is he a creature of pure contemporaneity—one of your *nouveau* New York pop-mod-hip fads. No indeed. For more so than is true of practically anything else, the hog, the *true* hog, as we shall see, belongs to the future!

III

Hog Husbandry

"If you're going to turn into a pig, my dear," said Alice, seriously, "I'll have nothing more to do with you. Mind now!"
—Alice's Adventures in Wonderland

Oftentimes it happens that the average citizen, in trying to get a proper handle on the hog so as to experience him in perspective, is sidetracked by considerations of the cow. Always there's the cow. Inevitable cow. We *think* cow. Cows are in our consciousness. "Cattle country" calls up instant visions of distant mountains and wind-swept plains leading off to nowhere and cattle grazing on slopes and tattooed men in wide-brim hats gathered around a fire with their horses standing stalwartly in the background; and one man will spit tobacco juice, and from somewhere off-camera, a cow will moo, very long and low, summing up in that one utterance its entire intellectual soul: "moo." And from that single word ("moo") we know of the presence of a cow and that God's presumably in his heaven and all's right with the world. And obviously, therefore, the men gathered here are cowboys or cowpunchers, or perhaps cattle rustlers; and here we are in cow country, and we all know where *that* is. (More about the cow later on.)

But who, among the teeming city masses, knows about "hog country"? Who knows where lies that land? Here we have all these hogs among us and hardly anyone even knows where they live. Nor who tends them. Cows we think about; hogs we repress. That's significant.

Another notable thing is this: Those who deal with cows are called "cowboys"; and of them a Texas rancher, quoted in *The Progressive Farmer*, declared, "Operators around here like to put on their cowboy boots and hats, get on a horse, and look over their livestock. They feel

46   THE HOG BOOK

downright uncomfortable around hogs." That's cowboys. The fellows who deal with hogs, however, are called "hogmen." Hog*men* and cow-*boys,* you see? That's significant too.

And what of the hogman? What of this lonely soul so much ignored in the public mind who labors out his days in an unknown land, a hazy netherworld called "hog country"?

Well, the first thing to keep in mind is that hogmen are not at all like ordinary men—not like them with respect to the fact that hogmen relate to, rub shoulders with and derive their chief income from hogs, whereas ordinary people generally do other things. Apart from that, of course, they *are* like the average individuals one sees every day.

In order to "talk hog" with a hogman you must first know how to say the word correctly and with the right inflection. This depends upon where you happen to be. Hog is not just "hog"; in fact it is almost never that way. In the South, it is "hawg" (like "dawg"); in certain Appalachian regions it is "howg"; in the middle and northeastern Atlantic states, and among Jewish hogmen (of whom there are very few), it is "haw-ug"; while in New England it is pronounced "hahg," and said with a faint grimace. Midwesterners, though—who have more hogs about them than any other Americans and who may, therefore, have some special insight into the subject—tend toward "hogg," with a very short *o*. In Ireland and much of Great Britain, hog is pronounced "pig."

Once you have mastered the matter of what regional dialect to use in speaking with your local hogman you will need to acquaint yourself with some of the more sophisticated hog terminology if you intend to carry on a conversation to any length. Obviously you can't go up to a Midwesterner, for example, and just say "hogg" and let it go at that.

To begin with, you will need to know that domestic hogs fall into several subcategories based upon sex and size, namely:

> BOAR–A male used for breeding purposes. Boars tend to be very proud and aloof, possibly because they are allowed a longer life span than the average hog, but can sometimes go impotent and display signs of acute anxiety.
>
> Sow–A female used for breeding. As a rule, sows are more gentle and courteous than boars, except when aroused by what seems a threat to their young, in which case sows have been known to fight off bears and wolves, and on at least one occasion—so cites Joseph Farqua in his epic tome *Sows Courageous*—a rabid caterpillar tractor. Otherwise, they are very motherly and affectionate and deserve our respect.

BARROW–A castrated, de-sexed male; a eunuch among hogs. Actually, most males are castrated before they are a week old.

GILT–A sexually inexperienced young female or virginal swinette. Once she has been mounted and bred, however, she loses this status and becomes a "sow." And everybody knows.

PIGLET or SUCKLING PIG–An infant hog from the time of birth up until about eight weeks of age, at which time he (or she) is weaned and usually weighs aroundabouts of thirty pounds.

SHOAT–A newly weaned youth (also called a "weaner") weighing between thirty and fifty-five pounds and generally regarded as a pre-adolescent.

YOUNG PIG or RUNNER–A hog more sizable than a shoat though not yet so large that you can glance at him just once and be able to say right off, "That's a *hog!*"

RIG–"We got a lot of terms in hogs that there's no particular definition for, we just understand one another," explains Blake Pullen, the highly respected Georgia hogman. "Like 'rig.' A rig is a male pig with one of his testicles up inside him. When you castrate him, you don't notice it and just leave it up in him, and he goes on the rest of his life tryin' to rub every gilt he comes to and just frothin' at the mouth and all, but he can't do nothing. This fellow who cut his testicles, see, he didn't pay much attention; he didn't count 'one-two.' I tell folks you don't have to have much sense to cut boars, you just count 'one-two.' If you can count to two, you're all right."

At the moment, there are 58.1 million domestic hogs being bred and raised on over 4 million farms by the good hogmen of America, most of whom are well versed in the ways of their own particular hogs and know how to relate and respond to them within a certain specialized frame of reference. We, as a people, do not (and perhaps *cannot*) know all there is to know about hogs and how best to deal with them. But through our anguished gropings and probings we are at least able, through trial and error, to discover a fleeting thing or two about how *not* to deal with them. The experience of the communal hogs at Koinonia Farm is an example of this.

Koinonia, ensconced deep in the rural bowels of southwest Georgia, is an interracial farming commune which was originally founded in 1942 to "bear witness as an integrated Christian community." Currently it serves as home for upward of thirty rigorously spiritual members who manage to keep the farm on a firm financial footing with the production of pecans, muscadines, corn, peanuts and pine trees. Back in the headier days of the civil rights movement, one of the main agricultural problems faced by Koinonia's black and white farmers was the constant threat of being blasted down in the fields by passing carloads of Kluxers armed with shotguns. But the farmers endured and persisted. Up until now. Now the problem is hogs.

Ed says it started about two months back, when an unexpected spell of rain in this normally arid territory soaked the peanut crop. The Koinonians harvested as much of the crop as was undamaged by the rain, but most of the sixty acres' worth of peanuts was left lying in the field, ruined. "Then someone read or heard from somebody else that if we just bought a batch of feeder pigs and turned them loose in the field they would eat the peanuts and at the same time fatten themselves up for market." Ed Lindsey is a tall twenty-four-year-old from the East who initially gravitated here for more or less philosophical reasons but presently finds himself a perplexed partner in Koinonia's hastily conceived hog-raising enterprise. (Feeder pigs, by the way, are young pigs purchased not for breeding but simply to raise until they're old enough for market.)

"Anyway," says Ed, "we bought two hundred and fifty small hogs to eat what was left of the peanut crop and turned them loose, and everything was goin' fine for a while until some of 'em just started dying." In about six weeks' time, fifteen porkers had given up the ghost for various causes. One reason for their untimely demise was found to be several opened-up bags of sulphur that had been left out in the field. Next, a few others began dying from some of the peanuts themselves, which by this time had commenced to rot. So it goes.

The sixty acres of ruined peanuts and pigs swoop across a fairly level expanse of land enclosed by a wire fence. One edge of the field borders on the part of the farm where the barn, tool shed and communal living quarters are located. Here, nearest the barn, Ed and his two partners have erected a wooden roof on posts to serve as a shady spot for the hogs. Nearby are two aluminum feeders, which are like upright rectangular bins with flat troughs extending out along the bottom of both sides. Each trough is covered by a row of small, hinged lids which the hogs must individually nose open to get to the food. The feeders were installed after it occurred to the farmers that a diet of all peanuts was nutritionally insufficient, if not fatal, and that their young pigs would require all manner of food supplements lest they begin passing away in even greater numbers.

Right at this particular moment a little cluster of hogs is scarfing up a load of corn from the feeders, amid sounds of smacking, munching, grunting and the opening and slamming of metal lids. Several more hogs are rooting around far out in the peanut field. Others are languishing in the shade of their shed, a few are lying in a tiny puddle beneath a water spigot, a few blissfully scratch their sides against the shed's wooden posts, while about a half-dozen more are sitting upright, frantically scraping their rumps against the ground. Their problem is worms—another aspect of hog husbandry the Koinonia hogmen hadn't counted on. Another was lice.

"We had a problem with lice a few weeks ago," recalls Ed, shrugging his lean shoulders with a slight wince like a mildly desolated scarecrow, "and then Millard told us that he'd heard the best thing for that was motor oil, burnt motor oil."

"How many pigs did that kill?"

"Two. We put the oil in little pits and let them wallow. A few had lice so bad we threw them in. Anyway, they'd wallow in it, and pretty soon some of them started drinking it. That's what did 'em in. It's always something like that that's happening. And when any single pig gets sick it's always, 'Ed, one of *your* pigs is sick,' and that kind of thing."

We walk across a piece of peanut field toward a former tool shed now used as a storage depot for a growing arsenal of hog supplies. "I put three sick ones in here this morning," says Ed, "while I went around and did other things—buried some other hogs—and those three got out and now I can't find them. You know, we got in this business for profit, but now it's just an educational experience." Ed opens the shed door and displays a recently deceased hog he intends to take to the local vet to determine whether it was done in by oil, sulphur, rotten peanuts or yet some other new type of peril. Right beside the shed a bevy of little hogs is browsing among the peanut plants. When Ed slams the door shut the hogs scatter in fright—all except for one that's lying close in an attitude of sleep. He looks to me as if he's dead, but Ed says, "No, not necessarily. They fake you out. You gotta go up and kick 'em." He goes up and kicks the hog, then declares, "Yeah, he's dead," and sighs. "Another way you can tell is if they're flat on one side." Ed shakes his head and lugs the newest dead hog to the shed to place alongside the next-to-newest dead hog.

The neophyte hogmen here at Koinonia bought their first 150 pigs at

$20 a head and then 100 more for $22. "With all the money we're losing in lost pigs and what we've spent for food supplements and equipment and what we paid for them to begin with," calculates Ed, "we'll be lucky if we make any money at all. Lucky if we break even. Then too, another thing that's discouraging about all this is that the *Farm Journal* and everybody else is sayin' don't go into pigs. We can maybe fatten these up to two hundred pounds. Today they're sellin' for sixteen cents a pound, but the prices are goin' down."

Ed walks on back to the feeding area, all angular and gangly with his shoulder-length hair flouncing out to the side with each lanky step, making him somehow seem altogether even more forlorn and put-upon than before: Sisyphus in the hoglot. There are about thirty hogs lounging around under the shelter, with most of the rest spread out across the landscape eating peanuts. Shortly, though, George and Chris, two of Ed's hog partners, bring the dinnertime sacks of shelled corn which they proceed to pour into the metal feeders with a sound that—though not at all loud—manages to arouse even those hogs from so far away they can barely be seen. And now they begin to converge. From every corner and crevasse of the field all the little pigs come scampering through the greenery, galloping across the peanut plants in their most excited stubby-legged lope, with their tiny ears flapping as they bounce and their eyes expectant and their tails a-twitter. It's a wave of onrushing joy, and it crashes upon the feeder bins with maximum ecstasy.

"Hey," says George, looking at the rump of a feeding pig, "I don't think this one's been castrated."

"Well for heaven's sake," pleads Ed, caught knee-deep in a moving sea of hogs, "don't do it now!"

The feeder lids have begun opening and slamming in a wild staccato that accompanies the frenetic noises of eating, plus the sporadic squeal of pigs chasing other pigs away from their private feeding sites. Pigs don't gulp their food as dogs do, but chew their meals more thoughtfully. "Eating like a pig" isn't necessarily a bad thing. Sometimes, of course, they'll submerge their faces right up to their eyeballs into whatever may be set before them—with all sorts of attendant snorts and slurps—but once they've got a mouthful, they conduct themselves in a relatively refined and sensible manner.

The pig-raising experience here, while it ranks as something of a disaster from both hog and human point of view, has at least provided the communal farmers with a whole series of enlightenments and minor epiphanies. "Pigs are really the cleanest animals you can buy," Ed declares with some degree of pleased amazement. "They just *look* dirty at times." George and Chris nod in agreement. "And another thing that's funny is the way they look out for one another. Like when one wants to scratch, another will let him scratch himself on him."

The hogmen here acknowledge that they still have an enormous amount left to learn about the creatures. Breeds, for example. The young partners in Koinonia's hog program don't have the remotest idea about what distinguishes one breed from another. In fact, they have no real way of telling the exact age or of determining anything distinctive even among their own small crew of hogs. All of the hogs here are named "Charlie." "Now that one over there," instructs hog photographer Al Clayton, nodding toward a little red barrow, "is a Duroc." Ed's face brightens. "Yeah? Where's he at?"

Then too, there are certain practices common among more experienced farmers that these urban-bred neo-hogmen here find difficult to accept. Ed, who used to work in a medical lab, says he was utterly aghast when the sixteen-year-old boy who delivered Koinonia's newly purchased load of hogs proceeded systematically to castrate all the tiny boars with a pocketknife, stopping every so often to wash the blood off his hands in a pool of pigshit and urine. "At least," Ed notes, "they don't bleed all that much. They walk funny afterward. But the *squealin'*. Lord. I remember Chris couldn't even eat lunch that day. We were havin' fish anyway."

There are actually two hog-raising projects going on at Koinonia at the moment. One, of course, is this large pig-and-ruined-peanut-crop endeavor, and the other is one conducted by a ten-year-old boy named Christopher, who has been raising four Durocs over the past year, ostensibly to sell but actually to play with. Christopher and Ed and Al and I stroll over to a small pen, well apart from the other hogs, where Christopher introduces us to his private stock and shows how they eat corn from his hand. "This one here," he says, climbing aboard, "gives you a real fun ride." And while he trots about the pen astride his Duroc, Ed observes, "You know, when they get that big they lose some of their angelic sweetness—if they ever had it."

It is late afternoon, on toward evening, and the entire stretch of peanut field and everything upon it is blurred golden and streaked out with long strange shadows. Some of the hogs lie together in clumps of five or six, facing into the sun, with their combined bodies seeming to meld more tightly together with the dying daylight into billowy mounds of continuous flesh. Most have come in from the field to gather in the cleared space around the feeder bins, and now either root alone or romp about with one another. The antics of small pigs are like the playful jostlings of little boys. They nuzzle and chase and even indulge in gleeful games of tag, all childlike and tremulous.

The rest root. When there is nothing else to do, or no remote threat on the horizon, a pig will root whatever patch of terrain he finds himself upon. A pig will root rocks, concrete, wood floors, anything. They are happiest, of course, rooting those patches of soil under which they seem to know they have the best chance of grubbing up something chewable. And they appear most pleased of all when rooting in concert with their companions: rooting and re-rooting and upturning everything and generally shifting the earth's surface. When actively rooting together like this they form a force like an ocean endlessly working over the landscape of a beach.

In time, the root-snorts grow softer. Small aggregations of pigs laze clustered in the dust and some in the little mud pool beneath the spigot. All are still but for the irregular twitter and flip of erect ears brushing away gnats. A few sleep. And of those who sleep, some snore and others kick or moan in response to whatever may be going on in their heads— for pigs dream, and occasionally even have nightmares. Apart from this, the only sound is the idle flap and clang of a few lids on the feeder bins: a noise which now and then will prompt some hitherto invisible pig to run stubbily in from the field trailing a bubbling stream of small grunts that sound like a man belching or clearing his throat.

Ed looks out across all these small precious pieces of pork, which now give him pause to ponder. He shakes his head: "There's always a certain tension about a bunch of pigs walking around. You never know when they're gonna flare up—start bitin' off another one's ear or something. You just don't get the calm, peaceful feeling like when you see a herd of sheep." He turns from the fence and begins stalking off, in his slow long-legged gait, to where supper's waiting. "We don't think we're doin' things exactly right," he shrugs, vaguely stupefied and exasperated, "but we hopefully'll be getting out of hogs for good—if we can just raise these we got now and get 'em sold off without any more major disasters."

West of here, another communal hog-raising experiment is growing into a down-to-earth economic success. But more than that, it's also an instance of employing hogs as a force for genuine social transformation. This is the Freedom Farms Cooperative, based in Sunflower County, Mississippi, and orchestrated by one Fannie Lou Hamer, who is black and tough and very wise and as deeply, fundamentally radical as a person can get.

The Freedom Hogs who wallow happily away on a newly purchased forty-four-acre farm comprise the co-op's "pig bank"—the first pig bank in the Mississippi Delta . . . or anywhere, so far as can be determined. "First, you see," says Fannie Lou, "we got donations and bought some land. Then the National Council of Negro Women sent a delegation down here and gave us fifty female pigs an' five males. What we do, we give the pigs out to needy families—they get one female each. Then they keep it fo' themselves, an' after they have a litter they give two pigs back to the pig bank. The idea, see, is to feed po' people—black and white. An' nothin' is gonna be sold until we take care of that need first."

Fannie Lou is an old-time civil rights battler who is sitting here now with a plate of chicken in her lap in her tiny brick house in Ruleville, surrounded by walls full of plaques, pictures of M. L. King, honorary doctorates and an engraved thing declaring Fannie Lou Hamer Day by the mayor. "The biggest thing that's been in the Delta has been cotton," she says. "What we plan to do is get enough land to grow our own vegetables an' raise our own pigs so we won't have to worry no mo' about our churrun bein' hungry. An' we gon' *stay* here," she slaps her knee. "We gon' make life better for ourse'fs, an' for the whites too. 'Cause no way they can hold us in the ditch without standin' down there with us." Fannie Lou smiles and nods like a sage, "You see, dey usta have dis ole starey hate look. But dey don't so much have it no mo'. Dese pigs is helpin' us toward real freedom, whether they know it or not."

*　　*　　*

Further inklings of the subtle and wonderful transformation hogs are able to effect upon society, and upon the human condition in general, are nowhere more evident than in the Midwest, the cornbelt-hogbelt heart of America. Twelve Midwestern states—Illinois, Indiana, Missouri, Minnesota, Nebraska, Ohio, Michigan, Wisconsin, Kansas, North and South Dakota and Iowa—serve as home for 77 per cent of American hogdom. And the people of this region seem honestly happier for it. Here, in this lush land of richest soil teeming with seas of ripened corn, fields of four-leaf clovers and amber waves of grain . . . here hog and man work proudly together in concert. It is not by mere chance that Virtue (as we are given to understand it) dwells in greatest proportions precisely upon that same span of soil where hogs thrive in greatest abundance. In Iowa, for example—where people brush their teeth eighteen times a day and read the Bible in the bathtub—there is approximately a full litter of pigs (eight being an average litter) for every single citizen. Iowa, with sixteen million hogs afoot, is the nation's richest state (in terms of hogs). There seems to be a direct correlation between this and the fact that one so rarely, if ever, reads news reports coming out of Iowa having to do with muggings, rapes, riots, rampaging hippies, gangland shoot-outs or other anti-social behavior. Nay, Iowa is veritably suffused with overwhelming wholesomeness. This same correlation holds true for most of the rest of the Midwest, wherein it appears that the beneficent principles of hogritude have pretty much permeated the public consciousness and smoothed out those raspier edges of society which are so apparent elsewhere. Now, inevitably—in attempting to rebut this self-evident parallel between the presence of hogs and the absence of human strife—someone will point to the Number Two hog-raising state (Illinois, with 7.2 million swine) and say, "What about Chicago?" Well, what *about* Chicago! Chicago just proves the point all the more. Chicago is the way it is because hogs have been ferried there *en masse* to be killed, not cared for: Chicago, the "hog butcher for the world," etc. Quite naturally this has an adverse effect upon the way people act.

In contrast to this, hogmen, and the families of hogmen, tend to be imbued with a certain serenity that somehow emanates from the hog itself. Hogmen are not all alike in their behavior, to be sure. But then neither are all hogs alike. Differences in the breeds with which different hogmen deal must certainly, therefore, account for something.

BREEDS OF SWINE

"Domestic breeds," proclaims the Encyclopaedia Britannica, "differ from their wild progenitors and even among themselves with regard to general conformation, colour, size, disposition and other characteristics. Breeds of pigs have been developed to meet certain geographical conditions and to satisfy particular economic wants."

As we all know by now, the old European farm pig was originally derived from the Wild Boar and was fed, housed and handled in such a way as to render him more meaty and less pugnacious. Serious *specialized* breeding, however, began in England. Some Siamese hogs were crossed with native stock and produced, in 1760, the White Yorkshire breed.[1] As the number of hogs in general increased, certain personal attributes began to stand out among those raised in one location or another until it got to where different hogs were becoming type-cast on sight. Then in the late eighteenth and into the nineteenth century people began categorizing and formalizing all these subtle nuances among the porcine race by establishing officialized pure breeds—thereby fragmenting the hog world even further and creating internal estrangement among its members. The U. S. Department of Agriculture advises rising young hogmen that "the selection of a breed is largely a matter of per-

Berkshire Gilt

sonal preference." From this it follows that a sociological probe into the dark question of who picks what and why—or what sort of people don't pick anything at all—may very well shed some light on the fine fix America finds itself in today! (Or maybe it won't.) Anyway, there are over three hundred breeds of hogs, and here are some that currently stand high in the hearts and minds of modern hogmen:

BERKSHIRE. Sometime around 1650, saith the legend, the Berkshire hog was discovered by Oliver Cromwell's army in the shire of Berks in England. Word spread of the "wonderful hogs of Berks," and in time the breed became a favorite with the snootier class of English farmers. The royal family kept a herd at Windsor Castle. In fact, a famous thousand-pound boar named Windsor Castle visited America in 1841, according to the American Berkshire Association, "creating a stir in the press which has seldom been equalled." The first Berkshires were imported in 1823. Berkshires are black with white points on their feet and the tips of their tails plus a splash of white in the face. Their chief peculiarity is a short upturned pug-like nose. Ears are erect and bent slightly forward; bodies

Hampshire Sow

Yorkshire Barrow

are solid and muscular. As for personality, the Berk is said to have a "splendid disposition," is alert, aloof and carries himself with great style. Most modern American hogs are descended largely from the Berkshire and other British breeds.

HAMPSHIRE. This is another import which originated in Hampshire County, England, and came to America in the mid-nineteenth century. The most striking Hampshire feature is the white belt that encircles its black body around the shoulders and including both forelegs. Hamps have straight snouts, narrow heads, trim jowls and erect ears. They are alert and active, and their sows are said to be "excellent mothers who suckle their pigs well." That says a lot right there.

YORKSHIRE. Yorkshires were developed, oddly enough, in Yorkshire County, England, and gravitated to the United States and Canada in the late 1880s, where they have since become among the most popular breeds. They are long, large and white (actually a sort of off-white), sometimes with black spots. They are prolific and notably pleasant to be around.

DUROC. The Duroc is red, nothing but red; sometimes even a bright cherry red. The breed was developed around New York and New Jersey

Duroc Boar

from hogs called Jersey Reds, who were also red. They were first bred by a man who owned a famous stallion named Duroc, whose name the local townsfolk also applied to his hogs. The Duroc is hearty, fast-growing, mild-mannered and reputedly brighter than other hogs (most of whom are also quite bright). *The Duroc News* claims, "The University of Kentucky over a five-year period of research proved pigs have high intelligence, superior to dogs, and have personal standards closely akin to humans. Durocs were better learners at all ages, achieving a learning level of 43 per cent compared with 22 per cent for Hampshires." (*The News* didn't disclose anything further on the question, "43 per cent of what?")

CHESTER WHITE. This breed was developed in Chester County, Pennsylvania, in the early nineteenth century and is found only in the United States. Some attribute its origin to stock brought over by William Penn. Chesters are large, prolific and all-over white. They are quiet, favorably disposed toward other creatures, easily adaptable to their surroundings, gentle and generally respected for their intelligence and sensitivity.

POLAND CHINA. The Poland China came from neither Poland nor China but—as you probably already guessed—from Ohio. (Poland China hogmen tend to get huffy about this.) The P.C. looks pretty much like the Berkshire except its ears droop down and its snout is straight. Poland Chinas are energetic and take plenty of exercise, which naturally helps them sleep better at night—as it would for all the rest of us. With their broad shoulders and well-rounded hams, Poland Chinas are quite popular among true hogophiles, and, in fact, they, together with Durocs, Yorkshires and Chester Whites, are the most widely distributed breed in the United States.

SPOTTED POLAND CHINA. The Spot looks just like a standard Poland but with white splotches on his coat, or sometimes Paisley shapes, or sometimes an outfit that looks as if it's been tie-dyed. On a very dark

Spotted Poland China Boar

THE HOG BOOK

night, however, you wouldn't be able to tell the standard P.C. from the psychedelic version if you met them on the sidewalk unless—aha!—you happened to feel the *ears,* which on the Spot are slightly larger. This breed is assumed to be a favorite among younger hogmen.

AMERICAN LANDRACE. This is a somewhat newfangled swine concocted in the 1950s from the Danish Landrace breed plus a few other things. The Landrace is the same shade of white as the Yorkshire but is longer, due perhaps to the fact that it has three more ribs than other hogs. It is prolific, fast-growing and, from all reports, appears happy.

Some of the older English breeds that haven't really caught on here include the Large Black, Middlesex, Essex, Wessex Saddleback, Suffolk and Tamworth. In recent years, certain innovative American hogmen have tampered with established breeds and come up with all sorts of *nouveau* versions of hog—each allegedly being somehow superior in this or that respect—but as of the moment none has become a real hit. I venture this may have to do with the generally drab-sounding quality of their names, things like Minnesota No. 1, Minnesota No. 2, Minnesota No. 3, Beltsville No. 1, Beltsville No. 2, Maryland No. 1, etc. It would be hard to get much worked up over hogs with names like these no matter what they could do. Now, the Montana No. 1 has recently been retitled the Hamprace—which, God knows, is an improvement—and, too, we now have the San Pierre and the Palouse; but thus far they've caused very little public stir.

Perhaps the major difference between American breeds and those from Europe is that the Americans tend to fatten faster. They are encouraged to do this. In fact, one of the uppermost duties of the American hog is to consume corn. Hogs provide the market for approximately half of the nation's gigantic annual corn crop—about 7.4 billion bushels. And in return, as if in ultimate payment for their keep, they are compelled to be transmogrified into more than half of the nation's standing supply of fresh meat. This is what it all comes to. Various breeds may shine forth in one regard or another, may be personable, courageous, droll, often profound in subtle philosophical ways, and certainly far brainier than most other domestic or even household beasts, but in the end they all are pork.

Most hogmen are uncomfortably aware of this internal contradiction that they not only live with but also help perpetuate. Naturally, modern hogmen know the economic intricacies of their swine-raising setups and usually have a refined comprehension of technique in terms of hog care. But most are also darkly cognizant of the fact that something more than mere meat makes up this animal. Something enigmatic, untouchable. These are men who observe each day—and voluntarily acknowledge to others—the intelligence of their hogs. They may even point out some of the surprising features of hog social behavior that they see. What they *sense*, though, they keep to themselves; and few among them fail to sense the vast, untapped potentialities of the hog, or fail to perceive the strange glow of awe that fills their own minds whenever they dare think about it. So they try not to think about it. Nor speak of it. But among themselves they know, and they know each other knows, and know each other knows that they know . . . and so on. Theirs is a state of almost constant tension between commercial and metaphysical considerations.

Indeed, among certain of the world's hogmen the meaning and spiritual value of the animal is too esoteric even to attempt to put into words. Only actions speak. As in the fall of 1970, when entire tribes of hogmen in the West Irian section of New Guinea went to war with one another over the killing of pigs. The tribe that lost the pigs took its rightful revenge among the members of another tribe. And then one thing led to another until they found themselves in a war that escalated to the point where ten thousand natives from twenty-nine different tribes, all armed with Stone Age weapons, were locked in combat purely on behalf of the hog. The war raged two months with scores of casualties before Indonesian authorities were able to arrange a truce among the tribal chiefs, who agreed to an armistice but warned—according to the Reuters report —"that if any more of their highly prized pigs were slaughtered, they would attack the police."

Among other of the world's hogmen above the Stone Age level, swine-raising may be considered an intellectually refined science, in terms of its technical aspects, yet it remains in essence a thing forever arcane and occult. In the Hungarian Gypsy tribe of the Kukuya, for example, hogmen employ a proven method of keeping hogs at home. "A hole is dug in the turf which is filled with salt and charcoal dust, which is covered with earth, and these words uttered—*'This is thine,/Come not to us!/I give thee what I can/O Spirit of earth, hear!/Let not the thief go!/We have three chains,/Three very good fairies/Who protect us.'* If the swine find the hole and root it up—as they will be tolerably certain to do owing to their fondness for salt and charcoal—they will not be stolen or run away."[2] As a specially sure-fire conjuration against possible hog rustlers, the hogman ". . . runs thrice, while quite naked, round the animal . . . he wishes to protect, and repeats at every turn: *'Oh, thief, do not go,/Further do not come!/Thy hands, thy feet/Shall decay/If thou takest this animal!'* "[3]

A cure for worms in swine, commonly used by itinerant Transylvanian hogmen, is to stand before sunrise in front of a nettle on which is poured the urine of the hog to be cured, while chanting:

"Good, good morrow!
I have much sorrow.
Worms are in my swine to-day.
And I say, to you I say,
Black are they or white or red
By to-morrow be they dead!"[4]

There are just scads of little locally-known things like this that could be enormously beneficial to all participants in the swine-raising world (not to speak of the animals themselves) if only international hogmen could convene perhaps a time or two simply to sit together for a spell and maybe whittle a little and swap hog jargon. More or less in order of their pig population, the principal hog nations are: Red China, Russia, the United States, Brazil, West Germany, Poland, East Germany, France, Mexico, the Philippines, Hungary and Yugoslavia. The very thought of all the myriad local theories plus pieces of private knowledge practiced among the *cognoscenti* in those lands is enough to bring a serious hogman to his knees in supplication. Language barriers mean little, for hogmen are more than simply colleagues in a common endeavor; they are initiates within an unofficial worldwide cult, into whose ranks only the most devoted enter . . . and none returns!

Though American swineherding has blossomed into a $5 billion Big Business enterprise, the basic working style of the good old down-home U.S. hogman hasn't succumbed to cybernation—particularly not when it comes right down to that crucial matter of the ultimate one-to-one, nose-to-snout confrontation. It is here, with all the trappings of civilization stripped away, that humanity and hogness converge most closely: a phenomenon which continually reinforces the transcendental character of basic hogmanism. The convergence of mind is perhaps best seen in the mystic American rite of hog-calling. For the call of the hog—or rather the call of the hogman calling *for* the hog—is one of those rare practices which, when done well, serve to reaffirm a man's primal sense of legitimacy upon the earth and kinship with other living creatures.

The art of hog-calling first emerged after it became obvious to those in livestock circles that, come dinnertime, the horses, cattle, sheep, goats and others of their ilk had to be sought out, rounded up and then driven to the appointed spot. Hogs, however, were always astute enough to recognize the sound of a human summons: not just the noise of any old someone idly yelling, but the *special* sound sent out by their own hog-man just for them. The most common of hog calls is "SOOO-o-oeyyy," which may be vocalized at whatever pitch or drawn out to whatever length one's particular hogs may feel comfortable with. In parts of the lower Midwest an extended "Who-o-eeyy" appears the preference of certain local pigs; while yet others sense something enticing in "HY-yee-aaa." Yankee farmers—generally devoid of the capacity for such free-hearted exhilaration as marks husbandmen out in the hinterland—make do with a dryly nasalfied "P-eee-agg, P-eee-agg."

This whole business became institutionalized as a rural and exclusively American art form in the 1930s, when the calling of hogs first took its place as a competitive event on those festive occasions when farm folk gathered for outdoor fairs. The relative artistic merit of each call has traditionally been rated by the judges on the basis of certain distinct criteria. First, the call must be strong enough to be discernible to the hog from behind a bush across a forty-acre clover field. Second, the quality of the call must be interpreted by the hog as one of warmth and friendly expectancy. Third, the call must be sufficiently original so that the hog can distinguish the style of the individual who's calling especially for *him*. Fourth, there must be a sense of persuasion and promise in the call, a flavor of charm compelling enough to woo the vagrant hog from his private digs. Finally, within the call itself there must emerge enough tonal variations so as to appeal to the hog's multi-faceted nature and changeable mood. In sum, as Fred Glanz, a onetime World's Champion Hog Caller, put it, "You've got to have appeal as well as power in your voice. You've got to convince the hogs you've got something for them."

The recent Woodruff County call-offs, which transpired just outside Fitzhugh, Arkansas, brought into the limelight a considerable amount of hog-calling talent, including one or perhaps two very capable good old boys you'll likely be hearing a lot more of in the days ahead. In contests of this sort, the hogs are usually just imaginary participants; emphasis is entirely upon the caller and his vocal and emotional range as rated by a panel of judges.

About fifteen hundred mostly local folk congregated in front of—but a respectful distance back from—the improvised stage, which had been fashioned from the back of a flatbed truck, and either stood or sat on the ground fanning themselves in the motionless ninety-four-degree Arkansas air. Places had been set up beside the stage for members of the Dixie Dew Licker Band & Marching Ensemble from Swint High School; but as it had taken longer than expected to get things ready, most of the players, in their regular heavy red uniforms with braid and epaulets and plumed shakos, trickled off one by one to lie in the shade. All that remained were a drummer and a cornet player who took it upon themselves to bleat a little fanfare preceding each contestant.

A Mr. Garvin P. Truman took first to the stage, to the strains of horn and drum, whereupon he bowed stiffly to the crowd's applause, cleared his throat and sucked in several deep breaths (microphones not being allowed in these events). On feeling the spirit, he aimed his chin skyward and let soar a fine, high *"WHOOoooo-oo-oo-yuh-up"* that floated wavering across the landscape and popped off with a quick whip, followed immediately by crowd whistles and claps and excited whammings of the drum. Garvin smiled, bowed again and declared, "H'it was my wife an' daughter what pushed me to comin' here. But I want you to know I've en-joyed it, an' I . . . an' I 'preciate it." Exit contestant. Another flowering of warm applause.

Next onstage hopped Clovis Combs, seventy-one, who claimed he learned to holler for hogs as a baby and had been doing it successfully ever since. Clovis, clutching the straps of his bib overalls, cut loose with a traditional *"SOOOO-o-o-o-eyyy, pee-yug pigpigpig . . ."* repeated over and over again with great flexings of tone, at times quivering like the mating *cri de coeur* of some deprived primate in extreme anxiety. On concluding this, Clovis gave an impromptu rendition of his "What a Friend We Have in Jesus" yell which had won him a bronze cup in Tennessee two years back. Then he broke into a brief clog dance, while simultaneously cawing like a crow, and finally descended the stage beaming and waving like a just-landed astronaut.

Little Jimmie Lee Rakestraw, twenty-two and noticeably shaking, took his stance in front of the audience and promptly appeared to forget what to do. Collecting himself, he raised his head and started in on a pseudo-Swiss yodeling sort of thing—*"YO-lo-lo-whaooo . . ."*—about midway through which he inadvertently lowered his eyes, caught sight again of the crowd, and in practically no time flat fell into a fit of nervous vomiting off the back edge of the truck.

At last up stepped Billy Ed Roebuck, probably the local favorite on account of the fact that everyone knew he was obliged to spend his days in a regular job-of-work in a wheel alignment shop and could get in his hog-calling practice only on a moonlighting basis. Anyhow, he sauntered to the center of the stage to the sound of applause plus the frail fanfare of a single cornet. (The drummer had dropped out by this time due to dizziness from the bald-scalding July heat, and had gone to lie down on the front seat of someone's car. Also, up until just before the last contestant, the five judges had insisted on wearing full-length black robes at their table fifty yards away, but they shucked them right after one judge dropped straight from his chair to the ground from sunstroke.)

Billy Ed stood stock still. Then, outspreading his arms like a night-club crooner, he flung forth full-mouthed a wildly abject tremulous ululation . . . *AHhhhh-oooo-eeeiii-ha-hah-hah* . . . with his voice climbing, gaining range and volume while concomitantly reflecting lush colorings of tone and shadings subtle as Bach:

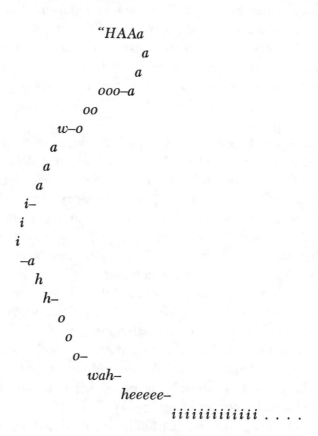

"HAAa
a
a
ooo–a
oo
w–o
a
a
a
i–
i
i
–a
h
h–
o
o
o–
wah–
heeeee–
iiiiiiiiiiii . . . .

It was a plorative unleashing of richest animal volume; a supersonic hog call; a call whose sheer piercing magnitude could break glass, puncture tires, separate tissue—Ahh, Ahh—a beckoning love song to swine yet unborn; a prolonged cry of evocation plummeting abruptly upward with highest velocity, beseeching the very heavens to send forth their wandering Prodigal Hog. And then . . . and then . . . with a quick shift in angle, it seemed to sail on up beyond the reach of human hearing altogether . . .

When Billy Ed arose from where he had momentarily collapsed, the crowd was still shrieking and applauding in rabid abandon. But more than that: From somewhere or other he had summoned with his call eight genuine hogs plus one impostor, two raccoons, five possums, a Dalmatian, a zebra, and somebody said later they thought they'd spotted a snake.

❀   ❀   ❀

If there are practices among hogmen that appear to overleap the bounds of "normal" human behavior or defy rational explanation it is because—and this must be remembered!—the hog *himself* defies explanation within the conventional restrictions of current thought. Naturally, then, those souls who are out there on the firing line, so to speak, who actually hobnob with the animals day by day, are the first to succumb to the romance of the hog—are the first to perceive their porcine charges as beings abstruse, exotical and everlastingly just out of mind's reach. Indeed, the common hoglot looms today as one of the last unexplored regions on earth, a realm known but to a few, and even by them not known very well.

"Research is the most badly needed thing in the hog business!" Blake Pullen, a portly, affable fellow in his late fifties, leans far back in his chair and swivels. "There's not much good research goin' on. There's just a lotta little ole common-sense things in hog production that you have to observe, things nobody don't tell you. You can't *prove* 'em. They don't put 'em in books." He stops and shrugs like W. C. Fields. "There are a lotta myths, too. Like water. These hogs don't need water to wallow in. They need plenty of *drinkin'* water. Hell, if a hog was supposed to be a water animal he'd have webbed feet. He wouldn't have his feet all split and forked up like he's got. What he needs is a place to be cool. He needs some shady spot he can get under and where wind can pass through. Pigs, in seekin' cool, have got good sense."

It used to be that hogs in America were divided into two classes, "lard" hogs and "bacon" hogs. Nowadays they're classified as meat-type and fat-type, with most farmers trying to raise the meat types, since there's no longer much market for animal lard. "Our greatest progress in hog breedin'," says Pullen, "has been in the last ten years. We've come up with a hog with a high percentage of lean and a low percentage of fat. Of course right now we're nearly 'bout up to our hips in hogs 'cause of overproduction. But there's still too many things we just plain don't know about 'em on account of the lack of proper research." As for projecting a shape for the hog of the future, the old hogman emeritus shakes his head, "If we change the models and styles of these hogs as much in the next ten years as we have in the last ten . . . Lord, I cain't even imagine *what* they'll look like!"

Clearly there's something about the old-fashioned raising of hogs—

some ineffable sense of fulfillment, some intimation of ecstasy, exaltation or whatever—that humans not only revel in but regard as eminently honorable. It has ever been thus. Historically, swineherding has ranked as a sacred trade. This is why, for example, Eumaios, Odysseus's old swineherd, who didn't like to sleep apart from his pigs, was deemed "godlike" by Homer. At bottom there's an innate animal kinship sensed here which the world's most dedicated hogmen come closest to bringing into fruition. Yet always offstage lurks the cold-eyed, all-consuming, cash-register rationale of an Established Society forever harnessed to earth by the blind autonomy of its own economic laws. Hence it is that U. S. Department of Agriculture Bulletin No. 1263 doth ordain: "The aim of hog-raising is the production of pork for human consumption. Every producer of swine should have as his objective the efficient and economical production of hogs that dress out high-quality carcasses."

Once the beginning hogman allows himself to be subjugated and sucked into this brazen frame of reference he finds he must relegate spiritual considerations to the most distal recesses of his brain and learn to concern himself instead with pedestrian questions of technology and technique. He becomes a cog in a machine, a faceless, standardized, predictable component whose chief function revolves around the eternal quest for the best technique for producing the most pork in the least time with the lowest outlay of overhead, labor, anguish etc. etc. Any concern he may feel for the simple happiness of his hogs must be reserved for his off-hours.

In terms of the priorities and practicalities of "commercial pork production," the farmer's first worry is how to retain the loyalty of his hogs—or, barring that, how to house and confine them so they can't just up and trot off to more exciting spots whenever the fancy strikes them. Experience has shown that plots of pasture, enclosed and made "hog-tight" with woven wire fences, are ideal, particularly since hogs not only can graze but also can provide for their own recreation and amusement. Hogs, however, being thick-skinned creatures, are highly prone to heat strokes. For this reason, the diligent hogman will erect what amounts to a little *ramada*—a flat roof resting on posts—under which his hogs can siesta in the shade. Then too, tradition (as well as the hog's passion for coolness) suggests there should be a comfortable wallow, preferably made of concrete and filled with a few inches of fresh water. Hogs prefer this kind of neat arrangement since they look most favorably upon clean water. Only if it's not provided will they resort to the discredited practice of mud-wallowing, which has already managed to give the hog such a bad name among city sophisticates.

Dreams of an everyday farmhog: Visions of sugarplums dance in the head of a mud-slumbering Yorkshire. Swine have intensely active, fertile minds; they just don't often seem that way. To you.

Alert hog guarding a tractor. Pigs are keen protectors of both people and property.

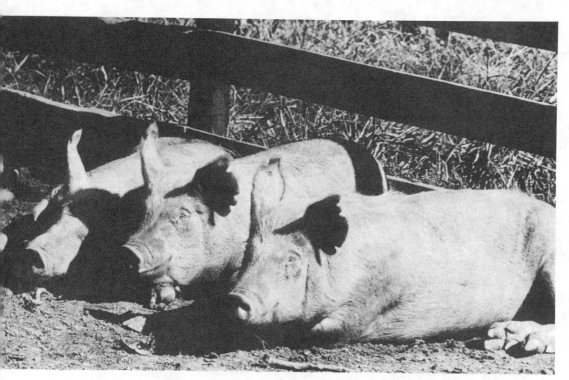

A trio of barrows basks in the pig-loving sun. Earnotches are cut for identification.

A serene scratching shoat at Koinonia Farm, a radical commune near Plains, Ga.

Whispering little porcine secrets nose to nose. In all of nature, hogsnouts are unique.

Catching his second wind, a Hampshire near Nashville readies himself for his next assault on this day's array of rootables. Hogs must root or die—as expressed in the old command, "Root, hog, or die!"

The main consideration, of course, is diet. Hogs don't necessarily eat more than other farm animals, but they do convert more of what they eat into meat. The hog stores 35 per cent of all he consumes, as compared with 11 per cent for sheep and cattle. Hogs are very good about eating: they'll eat almost anything, and if left to their own devices will supply themselves with a balanced diet. And, significantly, they also know when to quit, whereas certain other farm animals (horses for instance) have to be supervised lest they overindulge, bloat up and die. This is an aspect of their intelligence that makes hogs very rare in the animal kingdom; they don't overeat and can therefore be trusted. Hogs are so sensible about this they're commonly turned loose in fields to "hog down" (feed themselves on) corn or other crops meant for their consumption. Aside from corn, domestic hogs have a distinct fondness for wheat, oats, barley and red clover (which is mentioned here in case you have been considering gifts). Also, they take kindly to all sorts of commercial hogfeed. The slop-fed hog, though, is a critter of the past, since farmers are finding that a garbage diet not only maximizes the possibility of disease but also fails to supply all the nutrients a growing porker needs. Besides, when it comes down to a choice, no responsible hog deigns to grovel in a troughload of table scraps if there's a chance to scarf up a lip-smacking serving of some thoroughly modern hogfeed—as, for example, Wayne Gro-Pork 280, containing gustatory pleasures like bone meal, soybean meal, ground grain sorghum, animal fat (preserved with BHT), wheat middlings, phosphoric acid, activated animal sterol, riboflavin, niacin, calcium pantothenate, choline chloride, dicalcium phosphate, calcium carbonate, iron sulfate, manganous oxide, iron carbonate, copper oxide, zinc oxide, cobalt carbonate, potassium iodide, magnesium oxide, arsanilic acid and chlortetracycline. Now, the average human wouldn't know for sure whether this was a concoction designed to be eaten or detonated. But to a hog there's no question about it. From the time he's born up to when he's half a year old, the normal hog, eating stuff like this, increases his weight by more than 6,000 per cent. And the goal of the most avid pork producers is to make hogs gain this same amount in four or even three months.

Sheer snappety-snap assembly-line efficiency is the official virtue here, particularly from the viewpoint of those growing legions of large-scale hog moguls who think solely in terms of numbers—and who, in fact, may not be actual human hog farmers at all but giant "Agribusiness" consortiums run by air-conditioned accountants in silk ties, who operate out of unearthly places like Cincinnati. This is becoming the case in most fields of agriculture. The result is that the ordinary small family farmer—

the classical American agrarian, the proud pitchfork-waving straw-hat-ted shit-kicking no-collar hayseed who speaks utter ear-scalding scatol-ogy in King James syntax, who rails at citified debauchery and grudg-ingly rolls buxom milkmaids in the barn loft on days when no compliant sheep can be secured—is about to join the thinning ranks of wheel-wrights, coopers, office boys, bootblacks, sponge divers, telegraph mes-sengers, locomotive firemen and gargoyle carvers as a species slipping into extinction. In 1880, four fifths of America lived and thrived in the deep country. Now, what with mechanization, costs of fuel and equip-ment, U. S. Department of Agriculture policies that benefit large cor-porate-style farming enterprises, taxes and myriad other tugs and tam-perings from forces beyond human control, the active roster of American farms has dwindled from 6.6 million in 1920 to about 2.5 million hold-outs today. Of these, the hardest-core are the roughly half-million back-country operations that are still classed as "small family farms" (and that, under the present setup of things, are not long for this world).

So . . . with small farmsteads failing and folding up virtually by the hour in the face of competition from huge operators on vast tracts, and with future farm life projected in terms of ten-mile-long crop rows sowed and harvested by computer-controlled machines and scientists in heli-copters, what is the devoted man of the soil to do? The typical small farmer today is faced with the choice of fleeing to the city to fall into an essentially tragic and displaced sort of life . . . or of staying put to play—with a fierce, primitive determination—the very skillfully crafted and consciously contrived role of farmer *deluxe*. These men, you must under-stand, are a monumentally shrewd bunch of coots. They don't deck themselves out in those rancid overalls and chew tobacco and spit on their shoes and whittle and fart and talk about boring things in crude monosyllables and generally run the whole standardized gamut of prime-val farmer routines all because they just don't *know* any better. Ha. They know goddamn well what they're up to. And know exactly how they're acting. And what they look like. And smell like. And, above all else . . . they know it intimidates!

But the crucial thing to realize here is that really and truly they're not that way at all. For in the most private darks of their rooms—or among small groups of kinfolk or intimates—they unfurl their wings to reveal themselves as sensitive, scholarly or otherwise beautiful souls, many of whom may covertly write poetry, play the sitar, tap-dance or perhaps dabble in quantum mathematics. Yet their *public* stance is al-ways of purest aboriginal "farmer": yon mud-spattered bumbling coarse rube and everlasting follower of the plow. And why do they do this?

Well, it's all so very wry and sly and simple: If they can but convey the discombobulating impression that the way *they* are is the way *you* will have to be or the sort of person you would surely become if you took up farm life, then chances are you'll quickly choose to pursue some wholly different line and leave them alone in their bucolic peace. See?

It's the small-scale farmers, the fellows most threatened by corporatized agriculture and by the cities slowly closing in on them, who are most adept at role-playing. Now, I have been talking about farmers in general, but no less does everything here apply specifically to hogmen. As a matter of fact, hogmen have the poor public image of hogs as an added advantage in their favor. Whatever keeps encroachers at a distance is good for the cause. It has even been found on some farms that the hogman *as well as* his hogs play those classic roles they know will jar the senses of interlopers. Being "natural" is simply a pose; and it is becoming increasingly apparent that rural America is actually just a front for something.

THE STORY OF THE BARBED-WIRE HOG

Further explorations into hogness and hogmanism naturally led us to the small, old-fashioned swine-growing operation of a man who, out of respect for his privacy, we shall refer to as "Farmer X" (although his real name is Claude Hopgood and he lives on Route 2, Nebuchadnezzar, Tennessee). Farmer X "found himself" relatively early in life, and himself happened to be in hogs. He has "been in hogs" for nigh unto forty-four years now and is said usually to be willing, at the drop of nothing at all, to regale his neighbors and visitors with fascinating bits of true-to-life hoglore—most especially the wonderful tale of the barbed-wire hog.

Farmer X, on this particular day, sat on the front porch in his undershirt, sweating, scratching himself and, as he claimed, "jest out here listenin' to the corn grow." He carried a dip of snuff behind his lower lip and spat brown globs into the yard at proper intervals. Out in the yard, and all around the ramshackle house, wandered about two dozen hogs, who, almost precisely upon our arrival it seemed, commenced oinking obscenely and emitting foul odors. Farmer X rocked noisily in his chair and carved on a piece of wood. He was of medium height with weatherbeaten skin and scraggledy hair the color of sofa stuffing. He looked like a tree. The wooden piece he carved was to be part of a "better mousetrap" he was building so that, come this time next year, as he claimed, the world will have already beaten a path to his door, and gone on somewhere else. Above his door was a wood plaque with the words: "I plow deep while others sleep." With only the faintest nudge of persuasion, Farmer X began winding his way into the story of the barbed-wire hog.

"The Lord God, you talk about a stubborn hawg," he said in his flattened-out country drawl, "that 'un sure was." He paused and shook his head in admiration. "Like as if we didn't have woes enough already, what with machines an' one thaing or t'other." He peered out across the pasture. "Soon you gon' see them fields filled with robots—robots an' metal thaings sowin' an' reapin' an' plowin' behind robot mules with long antennas. Hard, even today, to find a real mule 'cause not much o' nobody's usin' 'em any more. But h'it's comin', h'it's comin'. You cain't prevail against it. I remember we was at a gospel tent meetin' back years ago when the first one of them big ol' combine machines appeared in the adjoinin' pasture, all wheezin' and belchin'. Nothin' like it had been seen 'round here before. An' the women screamed an' churrun commenced to cry an' run aroun'. An' the preacher broke out in Scripture-quotin' fits, an' then yelled out, 'On your knees, on your knees, ever'one, h'it's the spirit of our sins remembered!' Well, pretty soon it went away, and not till later did we find out what it was. But if you ask me, Second World's War was what done it. All started with that. I since seen many a God-fearin' hawgman come to grief pitted against all them modern de-vices made for growin' hawgs without ever touchin' 'em. Me, I like the way they look, I like the way they feel, I like . . ."

"How *do* they feel?" we asked.

"They feel . . . well, it's kinda like that old story about the elephant, you know. These four blind men were walkin' along and they bumped into an elephant. And I always think of that story."

"You mean a hog is like an elephant?"

"Well . . . its laigs are shorter. It's kinda hard for me to put into words —somethin' that . . . overwhelmin' . . . profound." He paused and his wizened eyes twinkled between thick folds of wrinkled farmerflesh. "Now these young college-educated hawg farmers, *they* can tell you—young college kids what come out after gettin' their master's degree in hogs. They can say. Like one of 'em, Walter Lonsdale down the road, he taught his hawgs to bring in the paper, watch TV, baby-sit . . . H'it's incredible th' number o' thaings they can do with those hawgs. An' I admire it, you understand. I don't understand it, but I admire it. Me an' my sons, now, we jest go in for traditional ol' hawg farmin'. Now when I was thirteen, fourteen year old, I admit, I'd sit out there for hour upon hour at a time an' I'd play the flute for 'em."

"And how did they respond?"

Farmer X shrugged, "They didn't care. Nowadays, though, I jest put 'em in a pen an' feed 'em and then do 'em in when the time comes. But it's not like I got no *feelin'* about it, naw. Hell, muh hands feel tired, back get to aching, 'round the small of it especially. This comes from a little thaing happened some time ago. I was out there mendin' some fence when this big ol' goat come up an' butts me in the back. Well, I guess he 'uz just tryin' to tell me somethin', but I didn't care, it hurt. So I went to the doctor in town. He's not really a doctor, he's a mule-collar salesman. Or usta be. Ol' Doc Wheatley. Anyhow, I went down for him to fix the misery in my back. An' I 'member he had this book there showin' all the parts of the body an' different parts of the backbone you can press to cure different ailments. You can cure cancer, you know, if'n you press the right spot. So he showed me on the chart where I got hit an' what was wrong with me. Accordin' to where I got hit by that goat, he said that shortly I would come down with rheumatism . . . or mumps. It was right between th' mumps spot an' th' rheumatism spot, an' he couldn't tell which it was 'cause he hadn't got that far in th' book yet. 'Cause you see it gets between them spots and, well, that's pretty compli- cated stuff. That calls for a specialist, and he warn't a specialist, he was jest what you'd call a 'general practitioner.' Of course, he wasn't really none of that, he was really just a mule-collar salesman by trainin'. But anyhow, he pokes me in th' back a little an' then winds up an' whops me hard as he could with an old Co-Colar bottle. Well, I jumped up an' said, 'Doc Wheatley, that 'uz the *dumbest* goddam thaing I ever seen in my life.' And he jest rared back laughin'. He said, 'Haw, I know it. I don't know nothin' about the back. I don't know nothin' about bein' butted by a goat!' Wellsir, I mean to tell you while he carried on like a fool laugh- ing, I picked up one of these old rattan chairs an' just whommed him over the head, laid him flat. As it turned out, my back got better later on that afternoon—which jest goes to show you don't need this modern med- icine after all. Nor streamlined farmin'. H'it ain't spirichual. These folks go on and on talkin' up the farm of the future an' all these fancy imple- ments. Boy, I tell you! What with th' way thaings change! When I 'uz comin' up, you'd spend all day choppin' and sawin' wood. Now they got these lightweight chain saws. Lorrrrd, you can saw, you can shore saw. Like I knowed a man, Wilbur Putnam it was, who sawed his whole house to pieces in the course of a hour! Didn't have no *earthly* idea what he was doin'. Just pulled th' cord an' didn't know how ta turn it off. Then one thaing led to t'other an' afore he knowed it he'd cut through the sup- port beams of th' roof. An' when that come down, th' saw just tuk off on its own. Went through th' floorboards, floorbeams, sawed up all the foun- dation of th' house until pretty soon th' en-tire structure come collapsin' on top him. We never retrieved enough of Wilbur to determine if'n it hurt him. I reckon, though, it hurt him. Reckon it did. Lord-a-mighty, I seen many a man come to grief from these mechanical contraptions . . . an' 'lectrical thaings an' gas thaings . . . Whole problem arose, way I see it, when folks turned to refrigerated air. We're jest losin' touch with our heritage . . . Th' old ways . . . H'it . . . h'it jest ain't spiritchual . . ."

Here his voice trailed off and his eyes took on the glazed-over thousand-yard stare of someone adrift in deepest rural reveries, all while his ropy old hands persisted whittling idly on the woodpiece for the better mousetrap. Clearly he wasn't of a mind to say anything more, nor to get into the tale of the barbed-wire hog—which, come to think of it, no one else had ever heard either.

Off to the side of the house, his two sons, Orvil and Eustace, were busily beating swords into plowshares. Meanwhile, Mrs. X was out back diligently locking the barn door after the horse had fled. And the hogs, for their part, continued to plod grossly around the barnyard with fulsome picturesqueness, rolling in dirt so as to appear all the more filthy, puffing out their cheeks to look as grotesquely bloated and "hoglike" as possible and enunciating the ugliest possible "oinks" with deliberate clarity. As soon as we had gotten down the road and out of sight, their sound changed gradually—changed then to one of something more like soft giggling laughter.

❖   ❖   ❖

Probably the most dramatic innovation in the world of the up-to-date hogman has been the "Pig Parlor." Pig parlors, essentially, are large, aluminum-roofed barns subdivided into small stalls with sloping concrete floors. The idea is to have everything under one roof. The typical parlor includes farrowing stalls, in which sows have their litters; nursery stalls for weaned piglets; and a series of "finishing pens," in each of which twenty or so young hogs are confined until time for market in a sanitary concrete setting supplied with self-feeder bins, automatic drinking fountains and spigots constantly spraying a fine mist of water both to keep the hogs cool and to wash away their leavings. The goal here is, as usual, efficiency: assembly-line swine. With a pig parlor (or "confined hog system," as it's also called), the hogman has better control over feeding, management and sanitation. Then too, claims one parlor owner, "hogs grow quicker with less feed 'cause they don't have to walk their feet off. Also, they don't pick up parasites." Yet for every gain in efficiency there's an equivalent loss in spirit. It becomes more and more difficult in such semi-sterile settings to view hogs as hogs rather than as neatly-packaged collected assortments of ambulatory pork.

This kind of social desensitization is even being encouraged in certain segments of the hog industry. The National Pork Producers Council—a federation of state-level groups banded together to boost the prosperity of the average pork producer—is currently working away, from their HQ in Des Moines, on propaganda efforts to disassociate hog and pork in the consumers' minds. It's a classic "Mad. Ave." press relations problem. Says the Council's program director, Craighton Knau: "The reason the hog got the image he did was due to man interjecting his lack of knowledge and his lack of management know-how. But you let a hog have his own way and he'll never create havoc in his own nest. He's as tidy as any human being you'd ever want to see. His personal habits are meticulous, believe me. Still, there's this stigma about associating the live animal with the finished product. It doesn't hurt a cow to give milk, but for some reason the housewife doesn't like to connect the quart of milk with the live cow. In our case, it's pretty fatal to the pig. And so, consequently, we've tried to divorce the two as much as possible. We don't think that as the housewife goes down that meat counter she ought to be thinking of those cute little pigs and looking at a pork roast."

What one envisions looking in the face of the everyday hog depends upon the angle from which he chooses to view. The same hog can be seen either in semi-abstract form as a strolling collection of pork cuts, or as an aesthetic and functional shape best suited to propagate more of his own kind. Now whereas the Pork Council, for one, wants to divert attention to the *product*, the various breed associations seek to rivet the public's eye upon the upright hog himself.

Along with almost every breed there exists an association of *aficionados* who make it their business to improve, promote and proselytize for their own special style of swine—often to the point of furious competition with rival bands of hogfreaks plugging other breeds. These breed associations are special-interest groups composed of the hogmen who raise the same kind of pigs. Among other things, the associations spread their singular versions of the hog gospel by means of magazine-style periodicals. Usually each such publication contains the latest factual tidbits about the breed, the minutes of a recent meeting, social notes, the president's exhortations to grow a better hog this year than last and—best of all—the services of outstanding individual boars, put forth in the form of paid ads. Here may be seen the very swarthiest of fine stud swine, all pictured with enormous testicles and rousing names such as Mr. Clean, Golden Stretch, Tarzan, Bonded Intruder, Score Again, Go-Boy, Sir Prize, Sir Snort, Hams-A-Poppin', Kem-O-Sabe, Mr. Standout, Sgt. Bilko and To-morrow: each one guaranteed to work wonders for whatever breed it may be part of. Some of the publications offering these pleasures are

American Hampshire Herdsman ("The December cover of the *Herdsman* pays special tribute to the great Hampshire females that continue to move the breed ahead"); *The Yorkshire Journal; The Duroc News; The American Landrace; Spotted News;* and *The Chester White Journal.* (One issue of *The Chester White Journal* contains an ad picturing a stud boar from the side, with the caption, "LUCKY LAD our son of Nugget in his working clothes." Opposite this appears a reminder of the upcoming "National Chester White Spotlight Show" in Athens, Georgia, concluding with the cry: "Start the Year Right with a Chester White.")

And now, here they are, here at last, hogmen together. Hogmen from Nebraska, Ohio and Texas; hogmen from Arkansas, Iowa, Kansas; from lowlands and highlands and deltas and mountains; hogmen in cowboy hats, jackets and Levi's: standing or strolling or talking in bunches or currying, clipping or feeding their porkers; boasting, joking, making deals among themselves, or bets. Hogmen all. All gathered in the Livestock Pavilion in Athens, Georgia, for that annual moment of potential glory, The Chester White Spotlight Show! (And in the small pens the quiet Chesters stand and gaze or lie snuggled on beds of straw or hold still while their masters dust their bright white bristles with talcum or Johnson's Baby Powder and paint gold numbers on their sides.)

There are 185 hogs here from fifteen states, all of them scrupulously groomed and prepared to parade their wares in the show ring. The two general categories of contests in shows like this are the competition and auctioning of top breeding stock, and carcass contests. In carcass shows, the hog is judged first by his appearance and demeanor in the ring and next, later, in disassembled form at the slaughterhouse. Some of the hogs here will win their blue ribbons in the packing plant forty miles away.

But regardless of whatever category they may be entered in, these show hogs receive the ultimate in human care. They are usually selected early in life and taught to respond to guidance with a walking cane. Their toenails are kept trim, their coats are brushed regularly to give "smoothness and bloom" and they are fed special diets. And now here they are: waiting in their private pens for the show to start while the hogmen and their families fuss over them, performing last-minute chores like trimming hair around tails, smoothing vagrant cowlicks ánd, here and there, touching up dark spots with white shoe polish.

Comes now the loudspeaker call: "CLASS ONE BOARS, CLASS ONE BOARS TO THE JUDGING AREA."

The judging area is an enclosed arena with a dirt floor and rows of bleachers on each side. At 1 P.M., the first wave of hogmen starts shooing Class One boars into the ring, whereupon a nervous few begin to sprint around on the dirt and root and drool and squeal. Others, though, instantly assume the mien of vain show hogs, and effortlessly, upon entering the arena, their workaday walk turns to a glamorous strut: all smug and white of coat, with snout held high. They know their way around. And all the while, the judge—a stocky, middle-aged man who continually puffs a cigar—is strolling slowly around the ring admiring porcine configurations.

Once into the arena, the exhibitors attempt to steer and cajole their hogs so as to show off their best parts and keep them from colliding or assaulting one another. Here, stepping deftly around the manure, are boys in denim jackets, young men with duck-ass hairdos, crew-cut codgers in overalls with brushes and bristle combs poking from their back pockets and their forearms white with talcum. A huge-hammed woman in a tight sweatshirt chases her sparkly white hog around the ring with a riding crop. Fretful men whap their hogs on the one side or the other with walking canes to steer them, each time knocking loose poofs of white powder. Every so often a hog will stop and shake his head, which quickly becomes lost inside a cloud of talc. The ring is a welter of men and boys plus a few women driving, tapping, whapping, shielding, stopping their hogs and all peeking up apprehensively toward the judge, who stands in the middle with his hands on his hips, squinting.

To the side, just past the arena's edge, two girls sit at a table selling Chester White mementos: Chester White T-shirts, Chester White ballpoint pens, tie tacks with a Chester White emblazoned in white on a black background, Chester White drinking mugs, decals, bracelet trinkets, on and on. And standing here, too, also hawking hogware, is none other than the National Pork Queen. That's right, THE National Pork Queen, standing right here, just like folks, wearing a rhinestone tiara and a green pantsuit with a broad sash running diagonally down the front saying "National Pork Queen," pinned at the shoulder with a "Hogs Are Beautiful" button.

The Queen's name is Marylin Bidner from Mahomet, Illinois, and she says she came because her official presence is expected, but also because her father raises Chesters and was going to be showing them here. Marylin, who is twenty, speaks in a voice soft as the down of a mouse, edged with a midwestern twang. She's eager to make converts for the hog cause. "In the Pork Queen Contest I spoke on why I thought hogs are beautiful," she smiles, and her tiara flashes in the arena lights. "I'd been brought up on a farm with hogs and . . . oh . . . the beauty I saw in them was the process of raising hogs and . . . and how the producers work to produce better hogs, and all the little things about 'em, like baby pigs and how cute they are." And did she ever raise a hog herself? "Oh yes, I had a market hog one time, a Hampshire barrow, that I was going to show in a carcass contest. And I walked him every day so he'd be tame to show. Then finally I remembered I had to drive him up into the truck that was going to take him to the slaughter plant. I wasn't too happy about that."

The Queen claims that early on in life she became aware of all the things the average citizen didn't know about pigs and pork—all of which ultimately induced her to try for the title. "I'd been around Pork Queens at the different fairs and all ever since I was little," she says. "I'd always looked up to the Pork Queens, I guess 'cause I was a little girl who'd naturally admire an older person of some esteem. But anyhow, I felt I could do something in this way to help the pork industry—help improve the image, you know. Like pork is often thought of as low-income meat. But . . . well . . . it's pretty difficult to change people's attitudes."

The judge is blaring into his microphone, "Let's give this class of boars a good hand, whadda ya say, folks." (Clap-clap-clap-clap.) "Thar's a lotta boars out here right now that can't help but do the Chester White breed a lotta good, I mean a *lotta* good." The judge now proceeds to discuss the individual pigs he picked, as each of the winners—from first to sixteenth place—is trotted one at a time across the length of the arena. "This boar that came in first has got integrity and muscle. He's got bone. He's clean. He'll sire some real good pigs . . ." Off goes Number One. "Now this pig that came in second is real well-balanced. Lotta correctness over that top, that shoulder. I'd like to see him walk a little lighter . . ." Etc.

After this there's the contest for Class Two boars. And now another phalanx of big snowy white hogs comes trooping into the ring with great *élan*. These hogs seem a little more skittery than the last batch. Hog No. 103 seems outright aggressive and lunges after the tail of every hog that comes into view, in spite of the frantic caning he gets from his handler. Occasionally, two passing hogs will snort and nip at one another, whoofing up foggy flurries of talcum powder. One eager-eyed farmer stands at ringside with a Polaroid, keeps urging his son to steer the family hog over toward the edge so's he can get a snapshot . . . closer . . . closer . . . AH!

Up in the grandstands, men in cowboy hats and pig tie tacks munch popcorn by the fistful. Wives talk quietly among themselves, nodding now and then in reference toward some particular pig, and sip Coca-Colas. In the row of seats nearest the ring, three guys laugh over a girl they know who's doing her best to navigate a hog around the ring. "Well, look a-there who's drivin' a boar!" one of them calls out. "Oh yeah," she snorts back, "*you* wanna come out here?" And suddenly her hog gets into a multi-hog traffic jam. "Shit," she mutters softly.

At moments, the arena is aswirl with hogs out of control: hogs running with thin streams of talcum powder trailing behind, men leaping over hogs, men cornered by hogs, men dragging pugnacious hogs by the ear. Abruptly, a boar will try to mount another, and the audience will laugh. "That's kind of a *feminine* squeal outta that 'un," one viewer yells. "I wonder if it's a real boar or not." And the audience laughs again. And the hogs snort and squeal. And minutes later a small portable bleacher with ten folks on it collapses, but no one's hurt and they all laugh. And everyone else laughs in relief, for they all care about each other. They radiate great warmth and rapport among themselves. They talk and feel on one another's wave length. And soon, after this judging, there'll be an auction of all the winners (some of whom will go for thousands of dollars), plus various private trades and transactions with multitudes of hogs changing hands among men from all different parts of the country who are melded together here in a common interest that fosters mutual trust and camaraderie. They're hogmen all. Keepers of the faith. Brothers in arms . . .

And after a while, somehow, one begins to sense the gentle, subterranean significance in all this. These living beings cavorting in the ring, occasionally leaving little deposits of themselves, are far more than they appear to the everyday citified eye: They are hog-and-man actively fused and functioning and *involved* together. And together they bear witness to the hog's power as a valid form of social currency and as a bond of human affection. These saintly white swine out here parading the richness of their natural wares on a dirt stage are a medium of understanding among people. For those people, the hog becomes a catalytic force that fosters creature compassion, a measure of selflessness, interpersonal cohesion, and the rare and beautiful sense of community which has begun to emerge as one of the highest forms of human knowledge.

IV
Hogs in Love

A Study in Sensitivity

Cool sows farrow more pigs.
—Ralston Purina Checkerboard
Serviceman's Pocket Manual

And then Moon began to feel a little fidgety in the limbs, just like the way he'd felt so many times before on overnight hauls when it would suddenly dawn upon him that he had been sitting in one position and not saying anything for a long time. In this case, he hadn't been keeping up his regular pace of banter and verbal folderol because, for a couple of hours now, Hoover had been slumped over sidewise, with his head against the rolled-up window, dozing.

Moon gave a sharp sigh. His left leg began mindlessly rocking side to side. He flexed his shoulders and pressed the accelerator a little harder to break the eerie steady rhythm of the road. He was restless and realized it; and now he realized he also had a mildly urgent need to relieve his bladder. His hands crawled idly around the rim of the steering wheel looking for something interesting, until one of them made an unexpected dive for the windshield wiper switch. The wipers came on with a whir, slapped a few noisy licks and then sat back down. But Hoover didn't budge. Moon faked a few coughs. Looked over. Still nothing.

Now, suddenly, his foot hit the brake for a split second and the truck jerked. "GREAT God-a-mighty," he cried, "D' jew see that?"

Hoover raised his head with a start and said, "Wha . . . ?"

"Biggest goddamn bear I ever hope to see! Almost jumped up on th' hood."

"What?"

"A bear!"

"Bear?"

"Boy, I reckon!" said Moon. "Or might even a-been a go-rilla. Happened so fast. You just better stay up and keep a watch out. No tellin' what-all kinda big thaings they got aroun' here that might run out on the road."

Now Hoover was sitting bolt upright—with the hazy, baffled look of someone who's been shocked wide-eyed—and said in awe to himself a few times softly, "Well, I'll be goddamned."

After a little vapor-hang of hesitation, Moon looked over out of the corner of his eye, then looked back to the road and whispered, "Faaan-tastic." Then louder: "Man, I tell you, that threw the fear in me so bad that I'm gon' have to pull over into the first place we find and answer the cravin'."

"Yeah, yeah, right," nodded Hoover, still piecing his head together, but now keeping his eyes fixed on the road shoulder where other big wild things must surely be lumbering along.

On up some miles more, a big, round GAS FOR LESS sign glowed from the roof of a little all-night filling station. Moon shifted down and maneuvered the truck off the highway, pulling up on one side of the station next to the door marked MEN. He put on the brake, leaned back from the wheel and stretched himself and climbed out of the cab, leaving the engine running. Hoover sat still for a second and then climbed out to visit the soft drink and candy dispensers. After a few more minutes he ambled over to the rest room, wherein he found Moon pondering an assortment of skinny vending machines affixed to the wall. This one read:

NEW!

Totally Different

CONFORM
The Revolutionary
Form-fitting
Prophylactic

Pre-shaped
to Fit Right
Gently Curved to
Hold in Position

Will Not Slip

Masculine Form
provides greater contact
for the utmost
in natural sensitivity
A NEW EXPERIENCE
in
*Comfort, Security,
Satisfaction.
Package of one
25¢*
Help Stamp Out **VD**!

Moon squinted in thought a second, then plunked his quarter into another device claiming to dispense "Instant Pussy"—and promptly received a little capsule containing a tiny plastic cut-out of a cat. "She-it," he snorted after a pause, feeling even a little more foolish than he'd expected, knowing that no matter what, he would never fail to fall for these things and would always walk away with the feeling of having been an accomplice to a rip-off that he deserved. "Hoover," he said, with a knowledgeable nodding of the head, "I see the fine Eye-talian hand of the Pope in all this."

The two laughed, then walked out into the bright night air and climbed back into the truck cab. Behind them the entire hog cargo—which had been fairly quiet for most of the trip thus far—was now up and grunting, shifting around and sounding distinctly sensual.

❋   ❋   ❋

Among hogs, sex is every bit as complex a social force to be reckoned with, speaking proportionately and with all things taken into account, as it is among humans. But with hogs it is far more open and forthright, and hence satisfying. Obviously contraception is of no consideration among hogs, but birth control, of an involuntary sort, does enter into the total scope of hog society by virtue of the fact that the majority of boars are castrated (and redesignated as barrows) usually by the time they're a week old. But those males as well as females who remain possessed of their natural physical capabilities and proclivities give ample testament to porcine sex life as a full-rounded phenomenon, frequently wrought with anguish and heartache, to be sure, but far more often touched with transcendent moments of shared tenderness and ultimate fulfillment. Indeed, those healthy, free-spirited and frequent assignations among them —openly courted and jointly partaken of—stand out as one of the rare genuine pleasures of being a hog. As Plutarch wrote in his *Morals:* "The greater part of men care not to converse with their own wives unless perfumed with precious ointments and odoriferous compositions, whereas sows attract the boars by their own proper scents, and smelling of pure dew, the meadows and the fresh grass they are incited to copulation out of pure affection; the females without the coyness of women or the practice of little frauds and fascinations to inflame the lust of their mates."[1]

Swine sexuality is a matter much pondered over, argued about, analyzed, assessed, evaluated and calibrated by men of science, and then rescrutinized all over again. The stumbling block in most studies of the hog sex act seems to occur when, just as things appear to be going well, academically speaking, the specter of intense joy is detected within the eyes or facial expressions of the participants. This usually messes everything up because it just doesn't compute in the context of what is supposed to be a "pure animal act."

There are, however, certain well-documented factual observations on which most swine scholars apparently now feel safe to agree. First of all, serious research undertaken by a team of professors and reported by the U. S. Department of Agriculture indicates, essentially, that the absolute numerical quantity of hogs upon any given geographical expanse *will not substantially increase* without the occurrence of reproductive activities.

Generally, boars attain puberty between five and eight months of age, while gilts first come into estrus (or heat) between their sixth and eighth month, after which they regularly come into heat every eighteen to twenty-one days. An analysis of the behavioral mechanisms of aroused hogs published by the University of Nebraska reveals the following: "The sexual behavior of pigs can be divided into four major areas: (1) The search for sexual partners; (2) the social contact between a male and female; (3) the pre-copulatory sequence leading to female's standing reaction; and (4) mounting and copulation.

"The interchange between male and female begins generally by naso-nasal or by naso-genital contact. When the male approaches, the female may run but he will follow the female persistently. During this activity the boar emits characteristic grunts, often grinds his teeth and foaming saliva surrounds the mouth . . . Sometimes a mock fight may occur. The male tries to bite the female's ears and neck, the intensity of the moment being completely different from an actual fight. The characteristic response of the female accepting mating is the standing reaction or mating stance. The sow stands immobile, arches her back, cocks the ears. When this reaction occurs, the male mounts and copulates quickly."

Now, there are lots more little notations and minute statistical items that researchers have compiled, but basically this is what transpires. It requires no external encouragement whatever. The role of the hog breeder, therefore, is simply to provide the setting and the circumstances at the proper time. But how does one *know* the proper time? "Well," as Blake Pullen puts it, "when she gets to frothin' at the mouth and he gets to frothin' at the mouth and it looks like they're gonna take the fence out between 'em, put her in the pen with him and in three minutes he'll be breedin' her. And when he gets through and comes completely down, take her out of that pen to where she can't even *see* that boar or hear him or nothin'. And in that way a boar can service a lot more sows. Just let him service her good one time and she's as pregnant as she would be if he lived with her for three solid days." In this way, Pullen says, "it's possible for one boar to service fifty females during a breedin' season without danger of overworking." Beyond that point, apparently, the average boar either loses some of his sexual libido or suffers from gonadal overtaxment or else finds that he hasn't sufficient time to eat or sleep properly.

Yet for all these numbers, the boar should not be regarded as some freebooting, swarthy profligate, as a "randy old boar." He is, in fact, of his own accord rather monogamous and loyal to the object of his affections. Says Pullen: "Now if you take him and turn him in the lot with a bunch of females—well, there's somethin' about nature. You turn that boar in with 'em, and one gilt comes in heat and he'll go to her off and on for two or three days. And there'll be other gilts comin' in heat, too, and they'll just be rootin' him and bumpin' him and knockin' him and everything else. But he won't pay no attention to 'em 'cause he just gets attached to this one gal, see, and he just keeps servicin' that gal, even though he'll have three or four others just a-bumpin' and knockin' him around."

Obviously, therefore, it is man who compels the boar to spread his sexuality so widely—and then man who also promotes the image of hogs as creatures colossally carnal and immoral. In Hong Kong it was the practice of the British government to award each Chinese widow with a sow, which would be bred regularly to produce pigs, and hence income. The British retained possession of the breeding boar and would have a soldier cart him, by bicycle, to call upon the various sows as they came in heat. This practice gave rise to lots of stories and little jokes about the attitude and facial expressions of the boar as he passed people on the street while being hauled off to service.

✧   ✧   ✧

This roadside sign says:

YORKSHIRES
Larger Litters   Less Feed
[And a picture of a Yorkshire]
"Most Popular Hog the World Over"
Houston White, Jr.

And on down another mile or so more, on the arid, sun-dazed outskirts of College Park, Georgia, lies White Acres, a lyrical, multi-chromatic blend of grassy pastureland backswept into majestic stands of southern slash pine, of fields and hills that roll and flow around a bluely tranquil lake, of barns and sheds and buildings, all showered in high sunshine, and of a white manor house enshrouded in the genteel shading of the chinaberry trees: all of it exuding a flavor of life as rapturous and softly pastoral as in those framed prints of impossible rural scenes found at Woolworth's.

Hogs grow here. Currently matriculating at White Acres are about seven hundred swine—consisting of breeding stock, which are generally free to roam within the sections of fenced pasture, and market hogs, most of whom are confined to the pig parlor. In the agricultural scheme of things the market hogs are the commercial point of the whole process, the product.

"You can get a lotta market hogs in a much smaller space than you can cattle," explains Houston White, Jr., who allows as to how he's been into hogs for about twenty years, "one reason being because they can eat ·concentrated food." And, as a matter of fact, the pigs in White's parlor are now doing just that: wolfing down high-powered corn and protein concentrates from metal feeders with lids along the bottom which they flip with their noses and then let fall, with a clang, when they're through.

This parlor is a sort of half-alfresco sort of thing, a long, low building with a metal roof and concrete walls that rise about waist-high so as to leave a gap of open space all around by which air can enter and circulate. Inside are a dozen ten-by-fifteen-foot stalls, side by side, with concrete floors and concrete partitions, also waist-high. Each stall, or "finishing pen," holds fifteen, twenty or so market hogs—all Yorkshires and all grouped according to age—who while away their days eating, keeping cool and relating to one another within the confines of whatever social hierarchy has been established in their separate stalls. The stalls are each equipped with self-feeder bins, automatic drinking fountains that supply drinking water when a pedal is pressed by a hog snout and "fogging nozzles" that spray a fine mist of water when it's hot, plus fans. The whole thing has the wet stony feel of a stadium locker room. And damned boring too—or so it would seem. "Yet," claims Houston White, "when you get right down to it, they have a pretty good life—just lay around and eat and be comfortable."

Though White is a generally non-emotional man, he seems, for the moment at least, vaguely vexed over his hog crop because, as he says, shrugging, "Prices are down bad. Down to the cost of production, near about." This is because the price of hogs depends on supply and demand. "It's not government-supported like other products are, mainly 'cause producers haven't wanted government messin' with us."

Rising and floating forth out of the several stalls is a mildly frantic cacophony from feeder bins slamming, hogs dimly murmuring, some squealing and some bunched together at the ends of their pens lowly snoring. They usually rest, play or snooze together like this at the ends where the feeders are located, while they deposit their leavings, very fastidiously and deliberately, at the opposite end, which White washes out every day. Right now he's hosing out the stalls and playfully dousing pigs who run delightedly squealing through the spray. "They love any kind of moisture 'cause they haven't got sweat glands, like people," he says, spraying. "If you had a clean swimming pool versus a mudhole I expect the hog'd go in the pool 'cause he likes to get in a lotta water—but of course naturally it wouldn't stay clean very long with his going in and out with mud on his feet."

A main problem here—common to most pig parlors—is keeping the confined hogs entertained, keeping them from being overwhelmed with ennui in the course of their concrete housed humdrum life as beings brought up to be eaten. In some of the stalls the little off-white hogs root and rolic with old auto tires. "Bowlin' balls are another thing folks use," says White. "They go to bowlin' alleys and buy old balls cheap. Lotta people pipe music into their pig pens, too, to keep 'em occupied with noise. Otherwise . . . well . . . you see that un back there?" He points out a certain doleful porker. "See, he hasn't got a tail. Got it bit off. That happens a lot. They just get bored, I guess. There's been a lotta research done on what causes this tail-bitin'. It's mostly overcrowding conditions—just always bein' in each other's way. They just get tired of it, and for something to do they start biting off each other's tails 'cause it's a convenient thing hanging down there to play with. See, here's another one without a tail.

"Another thing you have to contend with is that hogs have a social order, like people. If you put too many of 'em in one pen you have too many social problems, like with tail-bitin'. The thing is, though, it's just more *efficient* with these smaller units."

Here are these porkers gathered now in their concrete containers trying their best to acclimatize themselves to all this efficiency, striving to be properly sedate and modern. But they can't quite do it. When you walk alongside the row of stalls all the hogs scramble forward, bunching up against the concrete walls at the ends of their pens, all thrusting their snouts upward excitedly, with their nostrils flexing wildly away and sniffing, and the whole lot grunting in expectance, like a cellblock full of convicts waiting for their mail to be parceled out.

Houston White says people persist in imagining a hog as "this gigantic figure of an animal with gobs of fat, when the truth is that today's modern hog is a real meaty kind of animal, an animal that's got a lotta stretch to it, lotta red meat, real thick in the ham and a lotta muscling on the top." He points to one whose hind legs stand well apart when he walks on account of the good muscles along its inner thighs. This is called "wide tracking." And it is good. Altogether, White's pig parlor pigs qualify as the "modern meaty kind" whose physical virtues come partly from balanced, strictly controlled low-fat diets. "But your breedin' is something that goes right along with it—selecting animals for breeding purposes that have these desired qualities."

Breeding, however, or even casual sex, isn't within the remote scope of things ever to come for any of the hogs in total confinement here. I notice that of the males wide-tracking around the pens, nary a one is in possession of his family jewels. The males are all barrows—all castrated, all sterile. "Pigs aren't good to eat if they're left boars," White explains. "You can't sell 'em for market. In fact *these*," he says, pointing to three stalls-ful of five-month-old 210-pounders, "are ready to market now. I'll probably take 'em to the packing house and sell 'em this week."

Some of these hogs stand watching us over the walls of their stalls while others continue to jostle with their tire toys or scruffle around, vainly rooting the stone floor. *They* are what it's all about. They are the final fruition of the entire endeavor, the end result of human effort and hog reproductivity: the final arc of the circle.

But setting the circle in motion to begin with is the single perpetual function of yet other hogs carefully chosen and charged with the sacred role of reproducing the species in an orderly manner: boars and sows whose lives are prolonged for the single purpose (so far as they are concerned) of love-making.

It's a lesson that even the greenest hogman soon discovers, much to his amazement, namely: The stork doesn't bring pigs; *pigs* make pigs. And then those pigs make profit. And so on and so on. "We're all after the end product, which is a better meat hog," White says as we leave the parlor to go check on a day-old litter over in the farrowing barn. "But you never get the ideal. Who knows what the ideal is? The ideal meat hog would just be pork chop, bacon and ham. But you can't have that," he shrugs. "You've got to have a head and legs and tail and all of those things too."

The farrowing barn amounts to a sow maternity ward, a hog hatchery. It's partitioned off into separate stalls, some for sows in labor, some for just-weaned pigs, and special smaller enclosures wherein sows, flushed with the glow of motherhood, actually issue from their loins the Pigs of Tomorrow. Houston White stands at the end of a narrow farrowing stall and gazes down upon a portly sow lying on her side on straw nursing the piglets she produced the day before. Within the stall her own body is restrained by a framework of metal guardrails that prevent her from rolling over and accidentally smushing one of her young. "Now that un," says White, nodding toward the six-hundred-pound mother, "she was supposed to be the grand champion of some damn thing, but she hasn't had a decent litter till now." Clipped to the end of her stall is a small card containing the sow's name, boar's name, number and sex of the piglets and their date of birth. And here, reveling in the red glow of a heat lamp, are yesterday's pigs, eight of them, barely fuzzy, with wiry tails and soft ears and expressive faces already capable of smiling. Five have clamped themselves onto the available array of nipples which they suckle in rhythm, with their tiny heads bobbing steadily in unison, tossing up the row of sow teats on each pull, making pliable wet noises. Three others stagger bravely around the rest of the pen nosing things. They look like Chihuahuas.

Other farrowing stalls contain more mothers with different-aged litters, while in one lies a single potential mother seeming to be right at the edge of delivery. Meanwhile, in the next room, another prostrate pig-bloated brood sow languishes in a larger stall awaiting the onset of labor and emitting a litany of mournful oinks that echo deeply.

From when she first contracts pregnancy, a sow remains in the family way for about 114 days, at the end of which time she diligently erupts with an average of eight piglets, nurses them, weans them, rests a spell and then does it all over again. Most sows manage to produce two full litters a year and still somehow have time to tend to other things. This is fantastic fecundity, particularly when compared with the ewe, who takes five months to come forth with one or two lambs, or the cow, who usually produces one calf after nine months of being pregnant and irritable and generally getting on everyone's nerves. (More about the cow later on.) The hog, though, in the words of an 1855 swine treatise, ". . . is a perfect cosmopolite, adapting itself to almost every climate; increasing rapidly, being more prolific than any other domestic animal, with the exception of the rabbit; easily susceptible of improvement, and quickly attaining to maturity."[2]

Still, the subtle little nuances of hogritude are nowhere more evident than in the sexual patterns, protocols and procedures as they unfold and transpire between consenting adult swine—or "among," in the rare case of group sex.

Houston White claims his hogs prefer the morning hours as a time for conjugal conviviality. The boars, for one thing, just seem to be more in the mood in the morning, just after they've had breakfast and collected their thoughts and allowed their minds to drift off into erotic fantasies and pleasurable anticipations such as a randy little roll in the hay to start the day off right. The hogs referred to here are not the ones in the pig parlor but the older stock of breeding swine scattered around the pasture and beside the lake and in the fenced plots near the farrowing barn. One of these fenced sections contains seven young boars who this morning, apropos of nothing, achieved simultaneous arousal and mounted one another to form a brief, horny conga line of humping hogs—managing to hold this stance until one of them, or perhaps all in unison, decided it was an ultimately unsatisfying and profitless enterprise. Boars can, however, become impotent, and can even become homosexual, refusing to fraternize except with other boars.

Altogether, sexuality among swine is a very intricate and involved social function, carrying with it all manner of psychological pressures and expectations and potential anxiety reactions—occurring most particularly among boars. There are just so many little things that can go wrong. According to research on "Reproductive Efficiency of Swine," published by the University of Nebraska, males may suffer sexual maladjustment due to inadequate sperm, or because of having been "overused at an immature age." Or, they may have a "penis problem . . . due to an injury during copulation, or from rubbing the sow, hitting the ground with the penis, other sows or boars biting and fighting, or getting (it) caught in the fence or equipment." An additional possible drawback for a boar is "poor sex drive." Such a boar "may have a social problem through intimidation by a group of sows when first used, thus making him a shy or slow breeder. Also the boar may have been penned with another very dominant boar and only obtained an occasional chance to breed. The boar may have been raised under conditions where he never had the group learning experience of fighting, mounting, erection and general male developmental activity." The publication goes on to propose this: "It may be important to assist such a boar through at least one mating experience!"

Boars, of course, are not unmindful of their special, rather elite status in swinedom; nor do they neglect to comprehend that much is expected of them, procreatively speaking. A prodigious amount, in fact. Thus, with all things taken into account—with the boar's sense of duty to posterity, with his natural desire and need for affection, compounded further by his compulsive eagerness to demonstrate in the eyes of his co-hogs that he can rise to the present occasion, plus his keen awareness of the proud potent heritage of Boardom which he is obliged to uphold—with the sum total of all these responsibilities now riding, so to speak, on his gonads, he simply mustn't, mustn't malfunction. But, alas, many a stalwart boar has cracked. Now and again, under the accumulated strains of compulsory wholesale fatherhood, a well-meaning boar may wither limp right at the very brink of consummation. And then, if it turns out that he's unable to resurrect himself gracefully— either on the spot or for some time thereafter—he soon grows to realize he must henceforth face the awful reality of a sexual status far more dim and dismal even than that of a barrow: he is faced with being a *failure* as a hog. This is an awesome cross to bear. "It is hard," wrote Nathanael West, "to laugh at the need for beauty and romance."

On the basis of an extensive study of hog copulation conducted by swine scholars from France (naturally!), it appears that the female in heat ". . . assumes the major role in searching for the male, orienting the social contact in the sexual behavior sequence and stimulating the mounting reaction." This, of course, is enormously helpful for shy boars. But as farmer White observed, "When males get that urge, ain't no kind of fence will hold 'em hardly. They don't care *what* the sow looks like."

Right now he's attempting to round up eligible females to put in a pen with some boars (males and females being generally kept separate until breeding time) but can't seem to coax much cooperation out of them. "You gotta do it the way *they* want to do it," he declares. "You gotta use psychology on these things, and even *that* doesn't work part of the time."

At long last, hogman White gets his act together, with the result that now we find ourselves in a fenced swatch of hillside terrain, enclosed with a gaggle of aroused and tumescent swine, all presently engaged in preliminary investigations. From where I sit, practically the entire expanse of farmland can be seen, along with all the near and distant hogs who casually go on about their business beyond the confines of this particular pen: hogs on hillsides, hogs in the lake, hogs among the trees, hogs scratching, drinking, grazing, or mired in mud, hogs prancing in tall grass, hogs breaking wind, hogs sitting in the parlor, hogs asleep, awake, shaking their heads to the canvasy flapping of their ears, hogs lumbering, wallowing, trotting, grunting, gamboling or otherwise, in general, just living out their allotted span of hoglife to the fullest, maximal extent of its existability.

White Acres is located at quite nearly the precise point where planes making their approach into the Atlanta airport—ten or so miles north of here—begin their final descent. After a while, the jet drone and long contrails tracking across the sky become a part of everything else: the sun and clouds and birdsounds and low hog grufflings.

Here on the hillside, the boars and sows begin making their own approaches toward one another, each employing the regular courtship protocols recognized and accepted by all concerned. The boars begin their quick rhythmic snortings. This is a very distinct utterance employed for no other purpose. It's a burst of snorts that sounds very much like an idling Volkswagen engine, but deeper: "uh-uh-uh-uh-uh-uh-uh"—like that. While the boars carry on in this way (boars will be boars), the sows feign coyness, yet continue to move around the pen voluptuously, playing little eye games. Every now and then, both sets of contestants call time out to go get a drink of water. Then back to the main event. There are several males and females here who've been unexpectedly thrust together and seem, at times, to feel a little self-conscious about the whole affair. For the moment, at least, most of the hogs are acting kittenish. One sow responds to the hopeful onslaught of a courting boarfriend by sitting flat down on her rump like a dog. ("A clergyman who has bred many pigs," wrote Charles Darwin, "asserts that sows often reject one boar and immediately accept another."[3]) Occasionally a hog will drop out of the game in seeming disgust and go plop down for a therapeutic wallowing spell in a mud patch near the trough. Big jets trail new streaks across the sky, while here below these hogs root out neat nests of mud and pine straw and carefully settle down to contemplate their next sexual ploys.

There's a very aggressive small boar here who appears distinctly tawdry and boorish in the way he keeps scampering off in hot pursuit of every big-titted sow and gilt that comes into view. He seems to be acting out some personal boarhood crisis, a need to woo, to win, to dazzle and matter. None of them seems to want to have anything to do with him. But should a lone sow ever wander to within so much as ten feet of wherever he may be, he's sure to give chase—a brazen practice he persists in until, during one such romp, he runs smack headlong into . . . Constructor, the boar supreme. Constructor is large and long, with his shaggy bristles matted in red clay. He is a fearsome senior hog who speaks his remonstrances with great volume, activating the very air around him.

Constructor has been lolling in mud some distance from the rest, and only now does he commence to survey the scene. And, lo, what can that be which he doth espy beside yon tree, but a sow, smiling. And now it starts. Constructor stretches his neck, lowers his head and begins the swine "courting song," panting away at a furious pace. On being approached in this way, the sow's ears shoot up like spinnakers and her own grunts grow more frequent. And as the two converge, there begins now a series of surly and tender probes by both parties, though principally on the part of boar Constructor. They definitely communicate between themselves, touching noses to murmur intimacies in hogese which no one else can hear. Male and female move about, brushing sides, sniffing, poking and rooting each other in a dance of exquisite magic and sensuality. She, though of great girth, is nevertheless nubile and nimble. Her name is Sow 55, and she exudes a fetchingly serene charm as well as an ample bouquet of feminine smells best appreciated by her suitor. And he—ah, what a boar!—is the musky essence of style itself: debonair, dashing, full of ferocious gesture and grimacing, yet persistently convivial and possessed of winning ways, at times altering his velvety pitch to emit what seem like offhanded, devil-may-care oinks. Pure allure. Now he starts to rip the earth and paw the turf and grunt like a rogue. Still, for all his relentless vitality and raffishness, he comes across as distinctly gentle. Gentility is a personal characteristic among hogs, particularly in their intra-porcine relations, and most especially during periods of pre-coital foreplay.

Comes now a great trembling in the loins, an electrical moment of mutual knowledge. Suddenly she stands as if transfixed, all aglow in the height of amorosity, with her chin rising slowly and her ears limply bent backward. And he now, lifting his forelegs from the ground, rises, rises to rest his elbows upon her supple hips, and slowly leans forward, hunching over like one who's been carrying a great secret for a long time. His thin, vaguely corkscrew-shaped member is glimpsed only fleetingly before, like Cupid's arrow, it finds its sweet mark. Ho! On this note they begin to consummate their tryst. The scene dissolves to soft focus as Sow 55 affects a mien of uttermost beatitude, occasionally half-closing her eyes, while he, though completely motionless, becomes clearly lost in extreme exaltation, perhaps immersed in the long-held dream of fathering tall sons. The entire act of coupling and conjoining is completely internalized. Except for once when they take a few steps forward and a couple back, merely to maintain balance, they seem frozen: a study in technique and compressed passion. Their love-making consists mostly of the two standing like this, as if locked in concentration—which, according to the *Kama Sutra,* is the ultimate expression of physical-spiritual union. O burst into verse at the total tenderness, the trust, the intimacy! It is purest animal warmth openly shared in sensual sunlit air. (*Camera pans to swaying treetops, clouds, flowers, ants crawling on the ground.*) The act is not only mostly immobile, it is abstracted from all eternity. There is this sudden apprehension of timelessness in such intensified sexual communion, hog and human as well. Alexander the Great claimed it even frightened him into abstinence because it reminded him of death.

But surely physical love is the single most vital means nature provides of compensating for death, of proving you exist.

And now, judging by a slight shift, it appears that Constructor has reached his "little death." It is said boars have the longest ejaculation time among farm animals, lasting up to six minutes. They also secrete the largest volume of semen. Ah, but this is so clinical. It distracts from the deeply genuine sensitivity of the juncture now taking place—the essential sense of ongoing beauty and yet austere dignity so apparent in these loving creatures, both in their courtship and conjugality. Yea, mine eyes have seen the glory of the coming of the hog.

On concluding their session of passion, Sow 55 fairly floats on her trotters to the fence's edge and lowers herself down, first to her forelegs and then the rest following, her tail falling finally with a very neatly punctuated *plap* to the ground. Constructor, meanwhile, snorting in barely suppressed exhaustion, heads to the wallow, where he plonks down with a bulky wet—*fwash*—and coats himself in mud, rolling over and over and alternately rising to shake, with a loud flap of ears. In moments, he rises once more, all hideous with mud, to voice again his resonant call—"come, love spirit"—which rouses the sow for a few more mutual nuzzlings and sweetly muted grunts. They dote upon each other. He brushes hard against her side, making her muddy, too, and then glares around in protective defiance at the rest of the pen-mates—nearly all of whom snooze or lie in the sun, discreetly heedless of all that has transpired before them. At last, she lies back, tucking her legs under her, lies in the shade beside the fence with eyes closed. Now and again she gives a slight heave of a sigh and slumbers on. Constructor calmly scratches a ham against a fencepost before coming over to lie down at right angles to her, his head lightly leaning against her own. They are boar and sow fulfilled, their affection and sense of momentarily shared oneness having prevailed against all adversity. Hush. Let us sleep now . . .

HOGENESIS

And it came to pass in aroundabouts of three months, three weeks and three days, that Sow 55, far more girthy than before, being great with piglets, did alone seek lodging in a cold stone stall of the farrowing barn. *The Purina Hog Book* advises, "Be with your sows when they farrow." And true to this, Houston White is close at hand. Boar Constructor, however, he who compromised her virtue upon that barren hoglot, is nowhere to be found.

The lissome, winsome sow lies here on her side, with metal guardrails arresting her movements, and drones the low, throaty oinks of painful labor, though managing all the while to maintain a certain matronly lovelitude upon her face. And . . . now . . . a sharp *oink*. A tremulous jerk of her backsides. And out slides a fresh, wet piglet with his ears slicked back and his eyes closed and an umbilical cord still tethering him to his mother. Then—*oink*—another. And—*oink, oink*—still more, and an *oink-oink* here and an *oink-oink* there, with the first ones to emerge gradually detaching themselves from their cords and groping through the bedding of straw, eventually to discover Mother's udders. Through all of this, 55's face is perfectly seraphic, growing even more so as she nears the completion of delivery. It is utterly incredible that the coming together of two such shapes as sow and boar should ultimately result in a thing— let alone a whole set of things—shaped like hog. And here, at last, is the wonderful outcome: nine handsome piglets gleaming with life, averaging two and a half pounds each (two a little runtier than the rest), lying jammed together like cordwood, sleeping beneath the heat lamp which gives their natural Yorkshire whiteness a faint pink glow: tiny hogshapes in their first bloom of earthly being.

At this point, the hogman must perform certain vital functions of his own. First, he must clip and tie the navel cord of each pig. Then, with a pair of clippers, he shears off a set of sharp fangs called "needle teeth" or "wolf teeth," which would otherwise grow into tusks, or tushes, but which, at this stage of life, might do damage to the mother's teats. Next, the ears of each pig are notched in such a way as to identify the number assigned to his litter as well as his individual number within that litter. Finally, five days after delivery, the boar pigs not destined for breeding purposes are each laid upon a little device resembling a gynecologist's table and, with a couple of quick slits from a blade, out pop the testicles like two tiny California grapes. Hence are they castrated.

The pigs here—once they are fully warm, dry, de-toothed, notch-eared, castrated and medicated—will linger in this farrowing house (though in a larger stall) for their first several weeks of life, learning the basic ways of the hog world. Starting on the day they're born, pigs instinctively wobble to a far corner of the pen to perform their toilet functions. No other domestic animal is so innately housebroken. The sow, being a naturally good housekeeper, does this, too, thereby reinforcing the concept of cleanliness in the minds of her offspring.

The litter is weaned at about six weeks, up until which time they suckle up the six or eight pounds of milk per day that their mother secretes. This a communal sort of experience enjoyed by all. The piglets play in concert upon the sow's udders—which are sometimes known as. a "set of bagpipes"—as the mother oftentimes oinks out a sweet, cooing lullaby. Sow milk is full of protein and antibodies necessary for infant pigs—though, as Fred Seip of *The Progressive Farmer* notes, commercial milk sows wouldn't be practical because they'd require a suction device to encompass a whole line of twelve teats, and besides, the milk is "too thin for human tastes," or most humans, anyway.

Yet sows, for all their sweetness, are the very most unabashedly fierce and fearless of hogs to encounter if ever they feel their piglets are being threatened or imposed upon. Sometimes, it is true, sows will devour their young, but these are only the most neurotic mothers and this happens only rarely. Otherwise, though, there's little a mother can do to stave off the majority of worldly perils against which every young pig is pitted. The vast array of potential pig diseases alone, for example, is not only frightening to think about but often the ailments are hard even to pronounce. There's transmissible gastroenteritis, atrophic rhinitis, metritis, necrotic enteritis, erysipelas (which men can also get from handling the meat of diseased hogs), leptospirosis and salmonellosis. Then there's haemorrhagic septicemia ("blood poisoning"), pig scours (a potentially fatal form of diarrhea), baby pig anemia (an inborn malady), gut edema (a fatal intestinal infection) and brucellosis (incurable and also contagious to humans). Other prominent pig killers are pneumonia, tuberculosis, hog flu, hog cholera, ulcers, allergies and cardiovascular afflictions. Then come the worms: large roundworms, nodular worms, screwworms, whipworms, lungworms, kidney worms, threadworms, hog ascarids, intestinal flukes and trichinae. For sheer irritation's sake there are hog lice, ear ticks, fleas, flies and mange mites. None of the above, of course, excludes the concurrent possibility of fatality from yet other sources like predatory wolves, being struck by lightning or having a tree fall on top of a hog. If it's not one flirtation with disaster it's another. Everything is ominous.

"Garbage-fed hogs," says Houston White, "don't bring as much, 'cause they're not as good a meat. The law says you have to cook the garbage first to prevent disease." Trichinae worms are one of the problems hogs can pick up from garbage and then convey to humans in their meat. But the worst disease that can come from garbage (among other sources) is one that eradicates more swine than any other: hog cholera. This can wipe out an entire herd in the span of a few days. And if the disease itself doesn't kill an infected herd, state and federal agents will do it. Says Blake Pullen: "If a man gets hog cholera in his herd out there —and it's officially diagnosed as hog cholera—then you gotta go destroy the entire herd. Can't let nothin' live. You go out there and take bulldozers and dig holes in the ground, destroy that herd, then take all them hogs and put 'em in that hole and throw lime on 'em. Next, you have to take a steel rod and puncture every single hog, because after he's been in the ground a while he gets full of gas and, hell, he'll blow plum outta that ditch. The farmer, of course, is paid the full market value of those hogs, with the state and federal governments each payin' half the money." In states hit by cholera it commonly happens that multithousands of potentially afflicted hogs have to be executed this way and then burned to ashes or buried.

Now of course most of the death-dealing perils mentioned here apply to pigs who are already up and about. These are the lucky ones; for anywhere from a quarter to a third of all domestic pigs die almost at birth or within a day or two afterward, and nobody really knows why. Dr. Robert Seerley, a swine specialist at the University of Georgia, is exploring the question of infant mortality among domestic swine, versus the much lower death rate among newborn wild pigs. "A thicker hair-coat on the wild pigs is one reason they survive better," he says. "Faster metabolic rate is another thing, but we're still trying to find what else." Dr. Seerley had been experimentally crossbreeding domestic varieties with European wild pigs he's imported from the hills of North Carolina in hopes of hitting upon a hardier hog. "You see," he says, "the little domestic pig is attractive in terms of his hair coat. In the show ring they're attractive when they have this smooth hair coat—and generally that means a thinner hair coat. Through the years, in livestock shows, judges have simply selected animals that had less hair all the time. So more and more pigs are being born like human babies with very little protection. Of course, we're able with a human baby to take good care of him in a hospital. But out here in *these* conditions the pig doesn't get nearly the care." Dr. Seerley says that once all of the keys to the wild hog's survivability are determined, "we feel sure some of these things can be incorporated back into the domestic pig."

The interesting thing that emerges in all this is that in spite of how relatively frail, hairless and defenseless man has rendered domestic hogs over the years, they yet remain, all of them, completely interbreedable with their cousins, the Wild Boars. Thus, they could easily—and, within a few generations, completely—revert back to the natural attributes originally common to all hogs. But this doesn't seem to be the direction in which hogdom is headed. For the major thrust of the commercial pork-producing world today is toward ever more streamlined scientific swine management.

"The two most important factors in a successful hog enterprise," proclaims a typically dry-spirited swine scientist at the Connaught Medical Research Laboratories in Toronto, "are how many pigs you can produce and keep alive from one litter, and how long it takes to get a pig to market." With this as a goal, plus with the resources of (as well as mentalities attuned to) Technology at hand, teams of scientists, scholars and cybernetic experts have made great headway in the ultimate striving for pigs that can be produced as quickly and cleanly as so many computer print-outs. Both in the American Midwest and in Canada, these men, laboring away in their chalk-white deodorized disinfected thermostatic air-conditioned antiseptic lairs, have made such technological assaults upon traditional hogness as would cause the normal human mind to reel, if not revolt.

It's all in that hazy, semi-experimental phase for now, but basically it begins with inducing a prize boar to mount a dummy, wherein he deposits sufficient sperm to service as many as a dozen honest-to-god sows. But he doesn't get a chance to meet and mate with the real thing in a genuine carnal splurge. Nor do the sows (living in the super-sterilized Breeding Rooms of special Sow Houses) partake of his physical charms, except for those small portions of his precious bodily fluids by which they are artificially inseminated. No indeed. Boar and sow are mated by proxy, so to speak. The proxy being man. Then, after the sow duly gestates for the allotted time, she is ferried to a spotless delivery room, where she's laid upon a stainless steel table, strapped down and tranquilized. The table is hydraulically tilted sideways so that the sow hangs in a horizontal position. A sterile, insulated box with a door in it and rubber gloves dangling from two sides stands beside the table, into which go the piglets as they are removed by Caesarian operation. An assistant then inserts his hands into the rubber gloves and proceeds to cut and tie the umbilical cords and rub each newborn until it begins breathing. The pigs are next inoculated and carted off to incubators, where they are raised on artificial milk formulae, never to see or taste their actual mothers.

In this system, sows are viewed as simple pork machines and boars are vaguely undesirable characters who happen to make sperm. As for the pigs, the fruit of the act, life becomes a mere process controlled and calibrated up to the precise minute they hit the market weight of 210 pounds. They exist in small concrete cells with regulated temperatures and no light whatsoever—no sensation of day or night, hot, cold, or anything else to distract them from their uninterrupted few months of eating and enlarging themselves for the Apocalypse. It's an environment structured for total impersonality with the aim of turning out germ-free, computer-recorded pieces of living pigmeat. There's an odor of sanctity about the whole business.

The practitioners of scientific swinology eagerly espouse their new management practices with reams of properly academic jargon and propaganda technique. But, for all the pristine efficiency, it remains founded upon casual and amicably brutal assumptions: such as in the claim of one swinologist that his artificially sired, motherless, surgically delivered, bottle-fed superswine are creatures clearly unfit for the natural world because, "They're civilized; it's probably wrong to call them hogs at all. What they are, when you get right down to it, is pork." What all of this amounts to, when you get right down to it, is the ultimate example of sentient beings consigned to cipherdom: homogenized and fragmented, pigeonholed and put in a role.

But even greater feats of precision are already in the works. With the further perfection of artificial insemination techniques (the most recent breakthrough being the discovery of a way to freeze and store boar sperm), it's projected that, as one pork scientist exults with a zealot's gleam, "we'll soon be able to extract sperm from the finest boars and eggs from only the very best of sows, and place them together to fertilize, gestate and produce pigs straight out of a test tube." All of which, if you recall—harking back to the reincarnated remembrances of Joseph Myers—is exactly where pigs were said to have originated multi-millions of years ago in the lost land of Lemuria!

This kind of quest for instant canned hog is not, however, the exclusive purview of wild-eyed, abstracted woollyheads in white lab coats. Most of those men who regard themselves as modern pork producers actually hang out their tongues and pant at the luminescent, hopeful vision of a dreamy new hogfeed, or any other such innovation, that'll raise hogs twice as fast, and cut their lives in half. But "lives" in this context is a commercial term. Because the fascinating side point this brings up is that nobody knows for sure what the life span of a hog is anyway—that is to say, the natural span of years a domestic hog would live if left to his own. Some say twenty years, some claim he'd live as long as a man. But it's all speculation. Because the truth is, approximately 60 per cent are killed before they are six months old.

The old-style way of killing hogs on the farm traditionally takes place outdoors after the first frost of the season, the autumntime of gusting corn and the milling of sorghum. It is usually done in the morning. While water is being boiled in a metal trough or large barrel, the hog is knocked in the head with an ax or else shot just above the eyes in such a way as to stun but not kill him. Then a long knife is thrust into the hog's throat and twisted to sever the jugular vein. ("There is no cause to be squeamish about bleeding a live hog," notes a farm magazine, "because the sudden loss of blood produces shock, and the hog passes into a coma

in thirty to forty seconds. It may kick occasionally for two to five minutes
. . .") Next, the hog, now bled and dead, is dragged to the scalding barrel and soaked so his hairs can be scraped off easily. After this, the hog is
suspended by his back legs from a singletree, is split down the middle,
gutted ("use both hands to pull out the stomach and intestine"),
beheaded and then refrigerated overnight or left out in the cold to firm
the flesh for cutting and storing in the smokehouse the next day.

The age-old rural chore, of course, is vastly time-consuming and tiresome. But Technology and the machine-age food industry have graced
our lives in recent years with the modern packing house and the conven-
ience of delegated killing. Here it all amounts to the same thing, only
faster. Once in the factory, the hog is shooed onto a ramp, conveyed to
the "killing floor" where he's stunned insensible by electrodes, shackled
by a hind leg to an overhead conveyor chain, hoisted, jugulated, drained
and ferried to a long vat full of 180-degree water. On emerging, mechanical brushes whisk away all bristles, then vertical banks of blue flames,
between which the conveyor line passes, singe and harden the skin. Following this, the body is decapitated, slit down the center, disemboweled
and deftly stripped of all internal component parts (which are placed
upon separate conveyors), leaving only a carcass hanging from his hind
legs, bisected and opened up like a book. The pig's progress up to this
point has taken only a matter of minutes. The operating rate for a typical
packing house killing floor is about ten animals a minute—or six hundred extincted in the course of an hour, give or take a hog or two. From
here he is conveyed on along to refrigeration rooms to be more thoroughly fractionated into pork parts the following day. But nary a single
smidgen of carcass is scrapped or left uneaten or unused in some way.
The hog is totally recycled. Everything about him, in separate form, becomes ultimately included as an integral part of something else.

All of this is simply one last leg of the normal route to m'lady's lovely
serving dish. But, as the National Pork Council contends, it's something
probably best left ignored. We have become an urban race so blissfully
detached and compartmentalized that pigs are thought of as pudgy pink
things living in sties somewhere out in the boondocks, and pork is
regarded as a natural by-product of grocery stores that comes sliced,
bagged, canned or shrink-wrapped . . . and only the most distant connection, if any, is ever made between the two. Those who wish to enjoy
the illusions produced by scene painting and stage decorations should
surely never go backstage. Instead, think of pork: all that thiamin and
niacin and iron and tasty protein. Americans consume seventy-two
pounds of pork per capita per year, which comes to over 15 billion

pounds altogether, which is a far greater sum than the normal human mind is even able to conceive of. And just as well.

Or think of pigs. These pigs at White Acres. Sow 55's litter: newborn, squeaky-clean and lying here evenly heaped together under the red heat lamp, all gruffling, sniffling and making tinily muted pink squeals. And from time to time Mother, too, snorts out faint, pharyngeal noises whenever one of them occasionally chomps too hard on a nipple. Still, she senses, from having been through this before, that not too very long from now they'll be cleaved from her bosoms entirely and made into market hogs and fattened and foredoomed. She is operating under something of a deadline herself. (As Fred Seip says, "Sows are rarely kept longer than eight or nine litters. As they get older, their efficiency in terms of numbers of pigs they produce falls off. Also, they get so large they require too much maintenance.") Yet for the moment, at least, Sow 55 is aware only of this special set of saucy hoglets born today, each full of abundant animal spirits stirring now at her side, all precious and tenuous. "The most profound sentence ever written," claims a schoolboy in *A Portrait of the Artist as a Young Man*, ". . . is the sentence at the end of the zoology. Reproduction is the beginning of death."

Here they are, newly reproduced, roughly bright, wide-eyed and with wrinkled faces—squeal, nip, shove, suck—each one already fitted out with all the things inside that make a hog work. "In all animals," according to Aristotle, "at least in all the perfect kinds, there are two parts more essential than the rest, namely the part which serves for the ingestion of food, and the part which serves for the discharge of its residue. Intervening again between these two parts is invariably a third, in which is lodged the vital principle. There is a heart, then." And there's the awesome simplicity of a newborn oink. And somewhere within *that*, maybe—or floating in the air or else lurking in the earth, wallowing perhaps—is the possible answer to the question: Whatever makes that elemental vibrancy, the quality called life, awaken itself somehow inside a substance, shape, thing like a pig?

> *"So she set the little creature down, and felt quite relieved to see it trot away quietly into the wood. 'If it had grown up,' she said to herself, 'it would have made a dreadfully ugly child; but it makes a rather handsome pig I think—' And she began thinking over other children she knew, who might do very well as pigs . . . 'If one only knew the right way to change them—'"*
> From *Alice's Adventures in Wonderland*

V

Swine in Art, Sport and Show Biz

> Under your mighty ancestors, we pigs
> Were bless'd as nightingales on myrtle sprigs,
> Or grass-hoppers that live on noon-day dew
> And sung, old annals tell, as sweetly too . . .
> —"Oedipus Tyrannus," Shelley

Night circled. The road northward was mostly empty at this hour and so deeply backcountry quiet that there lingered the danger of being lulled asleep by the steady crisp clack of tires over the concrete joins. Only the rushing whine of passing cars (and this only at the most widely spaced intervals) managed to keep the two from slipping off hypnotically into highway trance.

Hoover, peering out his window at the roadside overgrowth and rolling seas of kudzu, suddenly—and with obvious relish—belched. He liked to do this, very loud and as drawn-out as possible, just to provoke people or sometimes to change the mood of whatever was happening or sometimes just for the hell of it. Anyway, it helped rouse Moon, who, feeling more alert, flexed his shoulders and began to waggle his cigar up and down between his teeth.

Then Hoover broke the silence again. "You know," he said, slapping his hands down hard on the tops of his thighs, "I bet you couldn't get twenty-five feet out into all them woods and vines without comin' up on a goddamn bear or some other man-eatin' thing."

"Aw, listen, man," shot back Moon, "I done come up on wild bears more times'n you got fingers and toes. They just as friendly and lovin' as critters can be."

"The hell you say!"

"Aw, yeah. An' the bears they grow around here in particular are goddamn *smart* sons-a-bitches too. Act just like folks."

"She-it," snorted Hoover, and slapped his thighs again.

"Lordgod," declared Moon, with a wink in his voice, "I seen bears doin' stuff you'd never dream of. I seen bears ride bicycles, roller-skate, tightrope-walk, all such as that. Hell, I remember this one circus act, too, I saw when I was little where this great big ol' brown bear and his trainer came out in the center ring an' just sat down at a little table with a bottle of liquor. Other acts were goin' on all around 'em, but they just sat facin' each other across this little table, drinkin'. The trainer, he'd pour some for hisself in a glass and some in a bowl for the bear, and they just sat right there and drank it like they was bein' in a bar. They both wore these businessman's hats, see, an' vests. Anyhow, they just sat there all while the other thaings went on—clowns an' horses an' acrobats an' such—an' didn't do nothin' but look at one another across the table and drink up the liquor. Then finally, after about forty-five minutes, the trainer fell out drunk and another man had to come lead the bear away. Most natural thing you ever did see," Moon nodded, with a sideward sliding of the eyes.

Hoover paused a second, belched very loudly in reply and then slumped back in his seat to watch the road flow by.

*   *   *

Actually, the bear, like most other animals that are called upon to perform, comes by his particular set of stage skills not so much because of innate talents or theatrical aspirations but as a result of sheer intimidation and duress applied to him. Dancing bears, for instance, aren't natural-born hoofers; they've been trained for this in the traditional manner, which consists of placing them upon a bed of burning coals or a red-hot metal grid, which consequently encourages them to "dance." The same is true for Tennessee Walking Horses, whose fancy high-step strut comes less from inborn skill than from sharp spikes driven into the soft parts of their front hoofs during training which make it painful for them to touch their feet to the ground.

Then too, there are seals that balance balls on their noses, kangaroos that box and tigers that jump through flaming hoops. But these are single-focus, specialized stunts which the particular animals initially have to be bribed or forcefully dragooned into performing. No one knows for sure why it is that animals do certain things or don't (like why hyenas laugh or how come whales don't get the bends), for animals have their own strange and secret ways that come across to most people only in terms of their most obvious individual or unique character traits. Eels are slippery, mice are quiet, monkeys have fun together in barrels, bats are blind, beavers are eager, geese are silly, larks are happy, cats wear pajamas, bears hug, rabbits punch, lions lie down with lambs, weasels go "pop," a worm will turn, wolves come to your door (unless you can keep them away) and horses are of a different color.

For the most part, these are aspects or activities that these particular animals, without any prior training or preparation, might naturally be expected to bring to the stage or screen if called upon. But when one attempts to tamper with an animal's talents or normal thought processes beyond certain limits all hell can break loose. One can never grasp, through a regular reading of your daily press, the full extent of the travesties commonly wrought upon animal entertainers, for newspapers omit, obfuscate and otherwise distort the shape of the real world almost as a matter of course. To get the unvarnished facts we rely upon the national sensation tabloids and the kinds of papers hairy boys sell on the street. One of these was quick to pick up a particular anti-animal article, which it carried under the headline: LSD-FED APE RAPES TV ACTRESS.

Now out of respect for the careers of the parties involved, I will necessarily omit names here. But it appears from all accounts that _____, an established professional actor and ape, was apparently slipped a hit of LSD by a cameraman while on location for *Beauty and One Beast Together*—a film is which _____ was cast as one of the principal performers. Once the chemical took effect, _____ freaked out completely, during the course of which he allegedly assaulted a Miss _____, said to be a rising young TV starlet. Miss _____ was consequently rushed to the county hospital for surgical repairs and ____ stitches. Meanwhile, the cameraman responsible was apprehended on charges of contributing to rape and mistreating animals. Production of the film was reportedly postponed until Miss _____ could recover adequately and the ape could "get his mind back together again."

Now hogs, by contrast, are just natural entertainers who manage to have their minds about them at all times and who, moreover, do not seem to require torture or chemicals in order to induce them to perform. On top of this, according to experienced animal trainers, hogs are far more versatile than most other animals, with talents extending over a wide range of theatrical activities. They possess, in fact, not only their own singular attributes but also the chief distinctions of many unrelated creatures. "The average hog," says one old circus hand, "can be as wise as an owl, stubborn as a mule, sly as a fox, quick as a wink, smart as a whip [this may refer either to whippets or whip snakes or to the U. S. Senate majority whip—it's that last one, I think] and, alas, poor as a church mouse."

The hog was one of the very first animals that man taught to do tricks. In France in the late fifteenth century, it was widely known that when King Louis XI fell ill there was nothing so effective for cheering him up as a troupe of pigs—all bedecked and festooned with pants and ribbons—who appeared before him to dance to bagpipe music.[1]

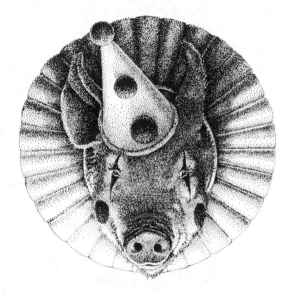

Another famous porcine entertainer was the "Learned Pig," first exhibited in London in 1789, who had been taught to pick up letters written upon pieces of card and to arrange them into words. "This pig, which indeed was a large unwieldy hog, gave great satisfaction to all who saw him and filled his trainer's pocket with money . . . One would not have thought that a hog had been an animal capable of learning; the fact, however, is another proof of what may be accomplished by assiduity."[2]

Similarly-schooled porkers were among the first circus performers in America too. Several toured the country in the 1840s with a showman named Dan Rice, who at one point when his circus wagons were bogged down, traveled up and down the pike exhibiting a trained pig at taverns and wagon stands. Countless circus clowns over the years have used comical pigs in their acts on account of the pig's keen sense of timing and generally convivial attitude. In the over-all context of show business, pigs are most often cast, in fact, as fleshy-nosed comedians, but they may also be called upon to dance, sing, do acrobatics or even play the straight man ("Mr. Bones") in vaudeville routines.

The fact is, far from being limited to the type-cast role of a gross, ungainly beast-form half-plunged in mud, the hog is a creature of remarkable breadth and subtlety. Taken out of their standard milieu, far from flyblown pigsties of yore, hogs almost invariably show themselves quite capable of super-porcine feats and amazing dramatic depths. Hogs adapt themselves with the tenacity of born troupers to stage, screen, circus ring, TV studio or to various dimensions of the communication arts in general. Obviously it could have only been through hogs' widespread associations with the world of the stage—and their eagerness displayed upon it—that the otherwise unlikely expression "ham actor" could have come about.

Swine-wise social commentators have noted, in recent times, how eminently understandable it is that hogs should take to the performing arts with such alacrity, as if propelled by some desperate eagerness to succeed. Hogs are not fools. They know the desolated, isolated, ghetto-flavored life of the hoglot is the only alternative available to most of them under the present scheme of things. Thus it appears that the standing, unspoken rule of life with hogs—as has been true with other closed-out beings and racial groups over the years—is: You seize upon what's open to you. And if the only option is to serve as the source of someone's entertainment, and if the show-business world is the only place outside hogs' standard milieu where they can gain some measure of social acceptance, then that's the route that must necessarily be pursued to the fullest measure of talent and energy.

In the 1920s and '30s, pigs were a prime feature in the exotic performances of Josephine Baker, a St. Louis-born black who allegedly introduced jazz to Europe and who, from one report, "set the continent afire with semi-nude gyrations to the beat of savage rhythms." Miss Baker, according to legend, "was toasted in the capitals of the world. Duels were fought over her and a sultan offered to dispose of his harem to obtain her favors."

Exotic dancers, of course, are not especially rare, but Josephine Baker's act was distinguished chiefly by the fact that she performed her gyrations amid a stageful of perfumed pigs who ran and writhed about in carefully choreographed accompaniment to all those savage rhythms. Out of appreciation for their talents, Josephine "walked her pigs proudly on the boulevards of Paris" and dined them in her home on rooster combs and champagne.

While Josephine's pigs were bedazzling the Continent, Fred's Pigs were already well-established performers in the United States. A review in the Salem (N.Y.) *Press* described them like this: ". . . From a clumsy, ignorant, peevish, man-cussed and heaven-forsaken beast, the magician hand of Fred Kerslake transforms the pig into an animated, clever, intelligent animal capable of competing in the show ring with any other creature that walks, crawls or wriggles."

One day in the early 1900s, so the story goes, Fred discovered twenty-one small pigs beside their dead mother. He proceeded to raise them, in the process of which he was (inevitably) struck by their intelligence. The next thing anyone knew, Fred had organized a vaudeville act: pigs on a seesaw, pigs on a ladder, pigs balancing balls on their snouts or riding around in an open carriage or pulling a pigcart—all part of the repertoire of a talented five-pig troupe that toured every major fair in America. Fred's Pigs went on to perform in Europe for ten years and even offered special shows for Kaiser Wilhelm in his palace before World War I. "Fred's Pigs," said the Roanoke (Va.) *Times*, "were well-trained and went through their stunts so naturally that the spectators were given the impression the pigs themselves had a well-developed sense of humor which they displayed in mirth-provoking pranks on their trainer."

Altogether, one of the main things Fred accomplished through his perambulations was to provide the Middle Americans of his era with fresh revelations into the porcine potential.

The most prominent pig ensemble to take up the cause after Fred's act is probably Uncle Heavy's Porkchop Revue. ("Think Pig!!!" says the brochure. "A ton of fun with Uncle Heavy's Porkchop Revue, featuring 'Oink,' the singing pig.") Uncle Heavy's act consists of three full-sized hogs plus one smaller hog, along with Heavy's fetchingly svelte wife, his little son Grumpy and Uncle himself—whose real name is Boyd Kimes.

Unce Heavy and his family and hogs headquarter themselves on a tiny farm in Neffs, Ohio, near the West Virginia line, whence they make their regular forays upon the sawdust trail, once even landing on the Johnny Carson show. Uncle Heavy, who is a gentle-spirited man, hugely fat and perpetually outfitted in overalls, claims, "I could be working right now if I wanted to. I went three years on the road in a school bus, but I got tired of it." Once, he says, he got his picture in *Life* magazine when he worked the North Carolina State Fair.

As for his hog talent: "Oh, one of them rolls out a carpet with his nose, another jumps a hurdle, another—see, it's a comedy act like a dog act, you know—another pig'll run around the front of the hurdle and miss it, you know, and another one'll knock it over and the other jumps over the top of it. Then they do things like run between my legs while I'm walking, seesaw, push a baby buggy. The tricks aren't sensational. I usta wear myself out to train 'em difficult tricks but the public wasn't aware of it."

The star swine is Oink, a five-hundred-pound Poland China who, shrugs Heavy, "will sing just whatever you want him to sing. It's a gimmick thing. You just touch him in different places and he makes noises. Like, I'll go, 'I'm Popeye the sail-or man,' an' he'll go, 'oink, oink.' And I'll sing 'Home on the Range,' and he'll go 'uhh-uhh-uhhhh-uh-uh,' like that. Then he'll go up the scale. Start out with a little bass grunt and go all the way to high C—startin' real low, you know—he'll sing, 'uuh, oh, ah, oOH, AHH, EHH, EEE . . .' and I say 'now hit high C,' and he goes 'ooof.' It's just an organized bunch of simple stuff, you know, but it's pleasing to people. I get by."

Uncle Heavy has been hauling his hog act thither and yon across the country since 1955. Before that: "I was in a circus doin' odd jobs and I wanted to be a performer. But I weighed three hundred fifty pounds and, heh, well, you know, it's kinda hard to place yourself. I had a dog. I trained dogs. But . . . I just didn't wanna fool with dogs. The thing of it is with hogs, though, is I *fit* with 'em. You know, like I say, I weigh three-hundred-and-somethin' pounds an' it doesn't look good to see someone who weighs all that much with a bunch of little dogs. With hogs, it fits. I'm just a hillbilly myself and we just buffoon it up."

When they're not performing, Heavy's hogs just amble around the property pursuing regular hoglike delights. "We don't have to rehearse very much," he says. "That's an interesting thing about pigs. You teach 'em once and they know what they're doin' from then on. Anyway, I got enough tricks in my act now where I don't need any more. So we never rehearse. Only time I rehearse is when I get a new pig."

Uncle Heavy's singing hog underscores a feature common to the porcine race as a whole: their fine lyric sense and general love of music. The pig's fondness for Terpsichorean revels has been noted and admired throughout history. As a traditional English rhyme puts it: "Come dance a jig, to my granny's pig, and pussy-cat shall crowdy" (i.e., play the crwth, or fiddle). Then too, in sculpture, paintings, ceramics, medieval manuscripts and wooden cathedral carvings, pigs are most frequently shown doing things like playing the bagpipes or sawing away on fiddles or plucking harps. "It has been argued that the pig as musician was intended to indicate that music was an occupation for vagabonds such as the strolling minstrels, but this argument loses much of its force from the fact that in the selfsame churches there are frequently figures of angels playing the same instruments."[3] The Pig-and-Whistle on inn signs or as the name for restaurants is another common association of hogs and musical instruments—and simultaneously a connection between hogs and the idea of being eaten. It's a fate that's difficult to get away from, even for pig performers who merit genuine star status.

One such porker who parlayed more or less standard pig talents into popular mass-audience appeal was Arnold Ziffel, a small Chester White who co-starred on CBS-TV's "Green Acres." Arnold was born on a farm in Indiana and didn't do much other than snooze, eat, loll around and perhaps endure the normal set of hog frustrations until he was discovered by Frank Inn, a Hollywood animal trainer, who provided him with the proper environment and outlet for his creative impulses.

Arnold first appeared on the show in 1965, where—again pointing up porcine musicality—he played the piano. As an actor, Arnold was also called upon to open doors, fetch letters from a mailbox, open a refrigerator and take food out, carry roller skates, schoolbooks and newspapers, pull a toy wagon, hold a pencil in his mouth and suck soft drinks through a straw. He was allegedly fond of Westerns on TV and could switch the set on or off as he desired.

"He was a wonderful contrast for Eva Gabor and fun to write for," recalled Jerry Sommers, the show's producer-writer. "Eva's an elegant beauty and with Arnold playing with her we got part of the old beauty and the beast identification in a humorous manner. In the show, she treated the pig just like a young boy and the audience loved it."

"Nobody cared if I'd powdered my nose," sniffed Eva. "When the pig was ready, we began shooting. In a sense it was most humiliating to share billing with a pig, but who wants to fight success?"

As a by-product of the rigors of his activity, Arnold garnered fan mail by the bushel, much of it from children (not yet hardened into the Establishmentarian outlook on hogs) describing him as "beautiful." He had fan clubs in his honor throughout the land, and, says CBS, more and more people are still reporting that he inspired them to raise pigs as pets. In 1968 and 1969, Arnold Ziffel was the winner of the much-coveted PATSY (Performing Animal Television Star of the Year) Award. And, being a star, Arnold was also touted about with all the cutesy drivel that goes into official network press releases, such as this one: "There's no business like SHOATbusiness if you happen to follow the antics of Arnold on CBS's 'Green Acres' . . . A pig is an unlikely animal to capture an audience's fancy, but Arnold seems to mean more to the TV audience than just another potential ham sandwich. According to studio spokesmen, Arnold is receiving many, you'll excuse the expression, 'ham' letters from children and adults alike. One 6th grade grammar school in Hilliard, Ohio, wrote fan letters *en masse* to Arnold and even promised to stop eating pork chops . . ."

Owing chiefly to his experiences with Arnold, Frank Inn has become a full convert to the hog cause. "There's lots of psychology in handling pigs," he says. "You can force a dog, a chimp or a horse to do something, but a pig, no. Pigs won't take punishment. Reprimanding will work with a dog, but with a pig, never. If you reprimand a pig he won't like you, won't respond to you and won't even take food from you. You can see temper in pigs. If I scold them, they scold right back. Still, I have trained pigs to do more than most people have trained dogs to do."

The show's director, Richard Bare, says that sometimes "Arnold did get temperamental—particularly in the late afternoons, like the other actors." Nevertheless, as Bare recalls, "We matched up a horse with Arnold in a two-shot sequence during one show, and it was no contest. Arnold zipped through his scenes, while we literally had to drag a performance out of the horse."

For the exercise of his showmanship, Arnold earned a cool $250 a day and was, moreover, protected by his union contract (Local 399 of the A.F. of L.), which guaranteed a minimum of $7.43 an hour with an eight-hour day and half-hour lunch breaks. But in spite of all his celebrity, he was not protected from the selfsame sure doom that would have befallen him had he remained totally sty-bound and undiscovered back in Indiana. The on-screen Arnold Ziffel was a perpetual ninety-pounder. As he aged and fattened, he gave way to one of his doubles, a younger understudy waiting in the wings. Arnold II similarly lasted a year and a half before being replaced by yet another double—in much the same way that Lassie, Rin-Tin-Tin and Roy Rogers' Trigger extended their careers long after their physical heydays. Yet once retired, *their* fates were more fortunate. As for the original Arnold, he was slaughtered, butchered and served up in pork chop form on Frank Inn's dinner table. But Inn didn't get to eat him. A phone call from the studio requesting the pig's appearance interrupted the dinner. Inn immediately had to go find another pig to train on a rush-order basis. The original Arnold Ziffel went into the freezer—and, when the electricity went off, spoiled. As a result, Inn has since never tried to dine upon any of his other trainees, vowing that he views that phone call as having been an omen from above. All these little epiphanies add up and come together.

Eventually—through accumulated demonstrations of porcine prowess, skill, sagacity, integrity and eagerness to please—humankind may very well be impelled to enter upon some over-all new awareness, some totally fresh level of consciousness and creature compassion concerning that ever-so-baffling enigma, the hog. Or again, maybe not—not yet. Nevertheless, it remains a sobering sensation to speculate upon and consider just how many undiscovered swine glooming out there in the boondocks presently project their lives beyond the confines of their pens or parlors and see themselves as zany hogs with gladsome faces, all gaily garbed in flarings and frilly rufflings, prancing gleefully beneath the klieg lights to the constant flowering of warm applause.

The hog is an experience in suppressed potential. The majority of them, of course, lie within their crude enclosures throughout the countryside fairly suffocating with unspoken, unheeded instincts and urges. Yet even so (in the eyes of those who are capable of seeing) they manifest an imperishable ability to juxtapose the most unlikely elements and hold them in suspension. They are self-willed and convivial. They are commonplace and shrouded in myth. They are quick of wit and without guile, existential and everlasting. And everlastingly fettered by secrets unshared.

THE HOG HUMILIATED

Nevertheless, beyond the scope of the professional entertainment world, the average layman's only contact with the live, up-and-about animal is likely to be under circumstances that degrade and belittle the hog, circumstances in which he is dealt with as an object of dumb mirth to be used and abused at will. The ancient English sport of "pig running" (said to be popular "at fairs and wakes") rates as one such typical debasement of hogritude. In this game, a large hog, soaped and greased and with part of its tail lopped off, was set loose before an onrushing crowd of hot-eyed young men, who chased and competed for the hog until it finally became the property of the lad who could catch and hoist it by its half-amputated tail above the height of his own head. Another fun thing was to put a small pig into a box having a drop-door, and to place that box at the edge of a large water tank. Lying across the tank was a greased pole which men were supposed to cross, then stoop and open the box. The fellow who could manage this feat won the pig. Usually, though, not only did the men fall into the tank but most of the pigs did, too, and drowned.

The American version of all this deliberate hog mortification is the "greased pig chase" (now mostly a field event for idle collegians), wherein a shoat, smeared with grease or motor oil, is chased, cornered, mobbed and groped at until the grease is either rubbed off or the pig is sufficiently worn out to be grabbed and held by someone.

All of the aforementioned put-upon pigs are, of course, relatively defenseless domestic varieties, and usually small ones at that. But then comes . . .

THE HOG REDEEMED

About A.D. 30, Strabo the Celt wrote: "Swine live abroad and are remarkable for their height, strength and swiftness—indeed it is as dangerous for the stranger to approach them as the wolf."[4] He was referring here to Wild Boars—both to those actually running wild and possibly also to those the Romans captured and raised for some of the violent and degenerate things they always seemed to be doing. "The Wild Boar formed part of the sports, pageants, and wild-beast shows and fights of the Romans. On the return of Severus from Arabia and Egypt, sixty Wild Boars fought each other; and in the year that Gordian the First was Aedile, he entertained the people of Rome, at his own expense . . . and one hundred and fifty Wild Boars were given out to be hunted, and became the property of whosoever was fortunate enough to catch them."[5]

As a rule, the average Roman shunned those sports, pleasures and pastimes that didn't have at least some small overtones of decadence about them, but for Wild Boars he made an exception. Boar-hunting was the *sine qua non* of courage, worthiness, virility and all the things that self-conscious warriors worry about. The Arch of Constantine pictures the potency of Emperor Trajan on horseback spearing a Wild Boar in the original style of the old Greeks. The truth of it was, the Romans really didn't add much to Boar-lore or to hunting technique itself, preferring instead simply to do it the way it had been done in ancient Greece.

Boar-bagging and like-minded rigors were much esteemed by the Greeks on account of these things providing splendid preparation of young people for warfare. In fact, it was Homer who recorded the world's first boar-hunt story:

> *Roused by the hounds' and hunters' mingled cries,*
> *The savage from his leafy shelter flies,*
> *With fiery glare his sanguine eyeballs shine*
> *And bristles high impale his horrid chine.*
> *Young Ithacus advanced, defies the foe,*
> *Poising his lifted lance in act to throw:*
> *The savage renders vain the wound decreed,*
> *And springs impetuous with opponent speed!*

His tusks oblique he aim'd, the knee to gore;
Aslope they glanced, the sinewy fibres tore,
And bar'd the bone: Ulysses undismayed
Soon with redoubled force the wound repaid;
To the right shoulder-joint the spear applied,
His further flank with streaming purple dyed;
On earth he rush'd with agonizing pain.

It should be clear that Wild Boars, be they alone or in groups, are by no means to be teased or messed with, taken lightly or regarded as anything less than the most formidable quarry men have pursued across the ages. Writing in the early fifteenth century, Edward, second duke of York, described the British Wild Boar: "Nor is there any beast that he could not slay sooner than that other beast could slay him, be they lion or leopard . . . And there is neither lion nor leopard that slayeth a man at one stroke as a boar doth, for they mostly kill with the raising of their claws and through biting, but the Wild Boar slayeth a man at one stroke as with a knife . . ."[6]

In India during the nineteenth century, the British, with their native fondness for almost any feats of derring-do that are vaguely pointless to begin with, took up the hazardous art of boar-hunting with horse and spear. "Pigsticking" became very much in vogue among British cavalry officers, who transmuted the whole thing into a ritual almost as stilted and structured as "riding to hounds" in pursuit of foxes. The idea was this: A team of hunters on horseback would trot through jungle thickets until they managed to scare up a Wild Boar—the presence of which was traditionally signaled by the first man to see it, who cried out, "Avast, there, a boar!" The others would instinctively know what he meant by this and would customarily reply, as one, "Huzzah! The very thing we're looking for!" On which note they would all give chase. For his part, the boar—having been rousted from his native haunts—would be understandably peeved and would dash off through the heaviest available undergrowth in a zigzag pattern simply to aggravate his pursuers. Eventually the boar might find himself ". . . crowded upon by several horsemen with spears, which they used in the manner of javelins . . . darting them into his body as they came up to him."[7] But not necessarily. "Hog hunting," averred one British pigsticker in 1867, "is not only more scientific, but it is also more dangerous sport than tiger shooting. If the horse is borne to earth in the charge, the rider will have little chance of escape." Indeed, provoked and agitated wild tuskers throughout the years have spilled the sweetbreads of many a good lad who simply knew no better than to intrude upon the sovereign terrain of a creature who was, and is, not only the most lethal quarry on earth but also the smartest.

All of this flash and peril just managed to render the boar business so much more romantic and so irresistibly British and foolhardy that in 1912 a wealthy American had a group of genuine European Wild Boars ferried over and deposited in the hilly reaches of North Carolina and Tennessee, there to be fruitful and to multiply. This is still the only section of the United States where Wild Boars abound. And it is here, too, that they hold forth, with all their desperate fierceness, against periodic onslaughts of hopeful big-game hunters. These boars have taken a natural liking to the region's dense thickets of mountain laurel and rhododendron, through which they can run for hours on end if necessary to elude dogs or hunters. Or they can spring from the bushes in a whirling flurry of razor-sharp tusks, if that's what seems called for.

Present-day rules declare dogs and guns to be the only permissible means of hunting boars—although there *was* a time (a single time) when local sportsmen and native farmer folk delved into the art of boar-bagging in a manner infinitely more exotic and flamboyant. The European Wild Boar had been imported from Germany to stock an enclosed fifteen-hundred-acre hunting preserve near the Tennessee-North Carolina line, where for almost a decade they were left totally alone to propagate. Then, in 1920, a band of local gallants, red of neck, mounted upon ponies or plow horses and brandishing spears, galloped into the hunting preserve to descend mightily upon this pampered crop of imported porkers—all within the highest traditions of British-style pigstickery. As the hunters clumsily advanced upon a covey of tuskers, the scene grew taut. Some of the nags began to balk. Some of the boars began to bristle. And suddenly, in a spontaneous eruption, about one hundred Wild Boars upped and charged topspeed right through the wire fence that enclosed the area and hotfooted off forever into the surrounding mountain countryside—also thereby closing, thenceforth and forevermore, the brief British pigsticking period in America.

Thus it is once more that the proud Wild Boar, through pure force of soul, redeems all hogdom from humiliation and restores respect among the species. "There is something noble in the courage of this powerful and solitary creature," wrote a nineteenth-century Englishman. "All his strength seems to be given him for self-defense. He injures no one, unless when disturbed in his retreat . . . He does not court a combat with enemies that thirst for his blood, but for the most part seeks to secure himself to the nearest covert. If attacked by savage dogs he rushes upon the foremost and strongest, maiming and killing numbers of the pack . . .

When wearied and tormented and forced at length to fight for his life, he turns on his persecutors and aims with vengeance."[8]

On another dimension, the hog is further redeemed by the fact of his presence or image (or visage) portrayed in an astounding amount of the world's great art. Now the hog, to be sure, is quite clearly a harmonious form, but this in itself doesn't entirely explain his attraction: It is that far beyond his physical symmetry he exudes, in great abundance, those tenebrous, enigmatic, ineffably archetypal aspects so instantly discernible to true artists. Hogs are depicted with dignity on ancient vases, amphoras, amulets, tapestries and stained glass; they come forth with galactic splendor in watercolor, oil, acrylic, pastel and ink; they are finely shaped into sculpture, wood carving, glass, pottery and all manner of precious stone. And finally, perhaps the most conclusive proof of how seriously the hog has been taken throughout man's artistic and cultural past is his featured appearance on monetary coins from places such as ancient Rome, Sicily, Barbados and Rhodes, as well as numerous other now defunct nations of the Hellenic period. A sow and a set of piglets are engraved on the Irish halfpenny of today.

"We proceed," proclaimed Aristotle, "to treat of animals, without omitting any member of the kingdom, however ignoble. For if some have no graces to charm the sense, yet even these, by disclosing to intellectual perception the artistic spirit that designed them, give immense pleasure to all who can trace links of causation, and are inclined to philosophy. We therefore must not recoil with childish aversion from the examination of the humbler animals . . . for each and all will reveal to us something natural and something beautiful."

Hog art began with primitive rock paintings, such as the well-known one of the charging boar at Altamira, Spain, but was expanded upon most notably by the ancient Chinese and the Egyptians, who enjoyed making hog amulets and small glazed pottery pigs. There's also a three-thousand-year-old toy pig from Egypt made of pure ebony with a lily on its forehead. The Greeks pursued the porcine aesthetic chiefly in the form of urns and vases picturing such scenes out of hoglore as Heracles fighting the Erymanthian Boar, and Odysseus' men changing into swine. The Romans translated their own hog legends into art, most commonly in the sculptured shape of the Great White Sow who, according to the *Aeneid*, marked the site on which Rome was to be founded.

Moving up through history, pig portraiture was eagerly indulged in by lots of the great masters, each of whom, in varying degrees, succumbed to what surely must have stood as the ultimate creative challenge for artists of their own time (as well as ours): namely, the quest to capture for all eternity the elusive, quintessential oinkness of the pig. Some of these endeavors include a copperplate etching by Dürer, a watercolor of the Wild Boar by Lucas Cranach the Elder, a pen and wash by Rubens, a hog etching by Rembrandt, drawings by Paulus Potter, paintings and etchings by George Morland and a lovely, moving portrait of "Cottage Girl with Pigs" by Gainsborough—all of which serve as even further evidence of the high place pigs held in the affections of old artists, and presumably of regular human beings as well.

In more contemporary times, Chagall burst forth with a painting of a green pig. Chagall was clearly possessed. In his autobiography he described his village in the calm of evening, while ". . . the white moon, enchanted, turns round behind the roofs, and only I, dreamily, remain in the same place. Before me, a transparent pig ecstatically buries its feet in the mud."

These artistic interpretations convey, in common, an over-all mellow, billowy feeling of hogness as a warm life-force . . . of Hog as a Primary Form suffered with benevolent and everlasting wholeness which rolls and flows across the mind, lovingly, and smoothes out unnecessary convolutions. Hog is together, artistically speaking.

A more recent artist to answer the grunt of the muses is Mr. C. W. Kello, who recently captured the "Best in Show" award at the Chesapeake Bay Art Association's seasonal exhibit. The winning watercolor represented three hogs sunning themselves on the bank of the James River. The judge said he selected the painting because of the hogs' stateliness: "The most elegant of the hogs is particularly dignified. The attitude of their bodies is one of repose and dignity, something which is difficult to pull off well."

A Richmond, Virginia, art critic declared, "The content of the painting is somewhat questionable." He preferred, instead, a seascape. But North Carolina's agriculture commissioner said, "I personally think Mr. Kello chose an excellent subject." And the secretary of the local Pork Producers' Association claimed he could rarely even look at a hog without getting emotional. "I just don't see how anybody could say hogs aren't an emotional subject. The hog was a friend to man before biblical times and arrived here before the Pilgrims."

Mr. Kello himself, as a result of his award, was swamped with calls from irate art lovers. "One lady phoned to say I should destroy my paint set. She was almost hysterical. She said the work was revolting. Another caller said that the painting smelled. I don't know whether that was a compliment or not." Though he's been offered substantial sums for it, Kello is determined to retain the painting because of his spiritual and emotional attachment to the subjects. "I found it ironic that such lovely animals could possess dignity while man was involved in maiming, killing and despoiling the earth."

More ecstatic still is "Pig," a four-by-seven-foot oil rendering of a 450-pound Yorkshire sow, in full life-size profile, by James Wyeth. The hog, whose name is Den-Den, was languishing away on a Pennsylvania farm when Wyeth decided to commit her to canvas. "I became purely enamored of her," he says. "There were good vibrations. Her eyes are so human too. Like a Kennedy's."

Shortly after she agreed to sit for the picture, Den-Den ate seventeen tubes of paint, a week after which there appeared rainbow droppings all over Wyeth's farm. The upshot of the endeavor, though, is a gleaming beige porker with a crooked grin. Another upshot is Jamie Wyeth's personal sense of new hog awareness. "I'd like to be able to have her in my living room—you can house-train pigs, you know," he says with a note of exciting discovery in his voice. "Also, I think the pig has a lot to do with what's going on today. I hope the picture is contemporary in all its ramifications. But I don't dictate what people should see and not see in it."

There is, Lord knows, a veritable plethora of things people see and don't see in pigs. Those artists and creative individuals who manage to open their eyes to the sunburst brightness of hogritude—and all that it connotes and conveys to us of the elemental urgencies of the earth—tend to glimpse such a shattering enormity of Truth and What's Real that they are usually forced to translate their perceptions into the form of fantasy, with hogs as the fantasy objects. George Orwell's *Animal Farm*, for one, handles hogs in terms of their intellectual eminence in the animal world and all the dark potential within them that lies moldering under their present set of circumstances. Read in one way, the book serves as a warning of the direct threat that overly frustrated and fed-up hogs could represent to man. As another example, in her play *Futz!*, Rochelle Owens makes hogs the means for presenting the mystery of man's eternal battle with instinctual and social forces. Her play deals with a simple farmer, Cy Futz, who is discovered to have been conducting a love affair with his sow, Amanda—a romantic entanglement of which the play's narrator says, "It's a pure sickness, but in its pureness it's a truth." At one point Cy confesses, "Sow, I know you love me but I wonder whether you'd rather be with your own kind? Piglets I can't give you, you know, though I am a healthy man." Nevertheless, he dreams on of their old age together when Amanda will sit "in my granny's rockin' chair readin' the Bible."

Hogs are commonly the focus for this kind of fantastic projection, both in art and real life. There was the fairly well-noted case in 1934 of a certain farmer named Ludlow Tharpe of Sibley, Iowa. Ludlow—being caught up, as he then was, in the desperation of the Great Depression— became convinced of the artistic possibilities of his livestock, particularly after he heard the Lord (or so he claims) speaking through the Gideon Bible, calling him to hogs. Suddenly it occurred to him that he'd be able to prosper on the profitable vaudeville and circus circuit if only he could hypnotize one of his brighter pigs and blot out her natural inhibitions about swinging on a trapeze. It is reported that he successfully hypnotized the creature, who happened to be a Duroc, but once she was induced into a state of hypnotic trance, as farmer Tharpe recalled quakingly, "That howg came forth with omens and revelations not found in the New Testament!" As a consequence, and with the encouragement of his neighbors, he slaughtered and ate the animal as quickly as possible.[9]

Still, for all that's been done to and made of the hog over the years in the name of art, he remains, at heart, not merely a performer nor an outlet for the expressions of other artists. The hog—even the commonest hog who struts and roots about—is an art form unto himself, to be properly responded to with awe and delight.

"The actual lines of a pig (I mean of a really fat pig) are among the loveliest and most luxuriant in nature," wrote G. K. Chesterton. "The pig

has the same great curves, swift and yet heavy, which we see in rushing water or in rolling cloud . . . There is no point of view from which a really corpulent pig is not full of sumptuous and satisfying curves. You can examine the pig from the top of an omnibus, from the top of the Monument, from a balloon, or an airship; and as long as he is visible he will be beautiful. In short, he has that fuller, subtler and more universal kind of shapeliness which the unthinking (gazing at pigs and distinguished journalists) mistake for a mere absence of shape."

The hog, indeed, is a creature of classic purity and directness, combining an exquisite joinery of leg and snout and bulk into a sculpted elegance shaped and polished by the elements, hewn, rubbed and burnished by wind, rain and sun. It is a truly transcendent and ethereal form, an elemental expression of higher life, or, in any case, of something far more weighty than we non-aesthetes imagine we perceive. Everything adds up. Perhaps, say some, the hog artfully mirrors the pathos of the country itself: huge, heroic, maladroit and always straining toward some elusive dream beneath yet another clod of dirt. Thus, both for now and for the foreseeable time to come, there is hope. For there is beauty in the beasts: there is True Art inside them.

❖   ❖   ❖

There was a pause.

Some of the hogs squawked and groped for their footing as the truck bounced over a rough railroad crossing; and Moon, who hadn't thought anything about them for quite some time now, began to ponder and speculate again. Then he said, "Really, man, why the hell do we hafta be doin' this a-tall? Why don't ya just let 'em go, let 'em run loose?"

And Hoover took in a deep breath, hesitated and, without turning his head, shrugged, "You just ain't supposta. Just ain't. That's all there is to it. Anyhow, they wouldn't have nowhere to go or nothin' they could do or no way to live without folks to feed 'em."

INTERLUDE
A Day in the Life . . .

They commenced to close in on me early in the morning: the Yorkshires first, then a few Hampshires, Durocs and mixed breeds; all of them making all sorts of soft, snorting, dog-sniff noises, with a few smacking their lips like a lapping sound. And they advance, now, up this little hill as if in a carefully coordinated assault by a squadron of Mongolian guerrillas specially schooled in stealth. Or like some 1950s Hollywood depiction of a gang of teen-age hoods (duck-ass hairdos, turned-up collars, short sleeves rolled high, Levi's low-slung and jack-boots and acne) gearing up to savage and pillage the clean-cut, crew-top, saddle-oxforded basketball star for Central High who slipped and squealed to the principal about how he had come upon Arnie, Rance and B.J. in the lower-level men's room smoking strange cigarettes and sniffing "that stuff." Ah, Lord, they're coming. They're gonna have at me for sure. Some of them even have rings in their noses and make low grunts as they move up, zeroing in. I feel certain one of them has a stiletto tucked behind his ham . . .

It began before daybreak here, all darkly gray and gloomful; nothing whatever to see except shadows and black heaped shapes. Then, around six, the grim light of dawn broke over the land, over the lot, over the nearest hog. And there he loomed: a gross, ungainly blob with no neck; all flecked and mottled with yesterday's mud. Not pink, but dirty black with a dirty white saddle that extended into dirty white legs; a tail not pert and coiled, but drooping straight down, flaccid and frazzled on the tip; not ears perked up erect, but fleshy things that drape frontward and flap like broken awnings. Here, in this place, gray-lit and ghastly, is a living creature of sorts—all bulk and snout and stubby trotter—thriving away in its normal habitat. And now it's come to this: I'm to be attacked and eaten by hogs.

This farm deep in middle Tennessee, Ed Hendricks' farm, extends across a stretch of rolling land just outside Brentwood. (These hogs are converging on me ever tighter: a relentless closing ring of hog upon hog upon hog, some pausing momentarily in their march to root up and chew a twig, nut, rock or some such, but most of them beaming straight in on me, moving with great furtiveness and yet steely determination. On they come: a few with drawn, lean faces; some jowly; some with ears wreathed in fringes of light-colored hair that gives their heads a quizzical, starburst quality. They surround me on this hill. There must be thousands of them—or at any rate, twenty or thirty . . .)

This is a small hill (or actually just an abrupt upcrop of dirt about twenty feet high) which overlooks a pond. Around the base of this hill stand five scrawny trees with no leaves. The whole sweep of this fenced area—three or four acres, I guess—is scattered with trees equally gaunt, while the ground itself is strewn with rocks and rippled with pocks, ruts, hog tracks and loose, rooted-up soil whose color runs from cadaverous orange to a dead sort of brown. (A trio of advancing hogs has stopped to poke and snuff in unison beneath a tiny ledge of dirt.) This is utterly denuded terrain. It is the grassless, graceless texture of bleak moonscape, or of wildly void waste and ruination.

The pond's edge borders along a wire fence, and beyond that fence— and parallel to it—is a gravel road. Running alongside the road are the L&N Railroad tracks and, past that, more trees. The trees sweep on outward until, in the distance, there vaults up the red-and-white WSM radio tower which sends the Grand Ole Opry electricitating out across all the national boondocks, twanging its way into thousands of little homesteads housing tens of thousands of farmers with virtually millions of other American hogs—none of whom happen to be closing in on me at this particular moment like the ones here.

A hog in the front rank steps forth and sniffs the corner of my boot-sole, sniffs and nips at it; and when I jiggle my foot he jerks back. Others press on toward the peak of this hill. I am sitting on a metal something-or-other—I think it's the top grate of an old gas space heater. A hog bites at the edge of it and I hear his teeth clank down on the steel-hard surface.

The swine who've come upon me in this sunwaked morning cool are all somewhere around three to four months old. A black Spotted Poland China with a long, sallow snoot has just nibbled the elbow of my jacket. All at once, the rest, forming a perfect circle around me, begin snuffing and thrusting their snouts forward, with their little dark eyes probing me in wonder, and a single foreleg of each cautiously poised as if in mid-lope. Then—after several frozen minutes—a few turn back, evidently satisfied, and proceed to root studiously at the earth with an apparent sense of reassured safety. I feel the same way.

Every twenty minutes or so, big Mack trucks heaped with dirt drone down the gravel road, at which time most of the hogs scattered around the lot look up and snort or else just stare and say nothing—think about it, maybe. In this brighter morning light, they seem much less ugly; their bodies appear quite functional in their design, actually. Or in any case, certainly not crude.

There are snarls of fallen limbs, slivers of board and wrenched bits of corrugated iron cluttered all around the lot. I'm sure every bare extrusion above the surface and every square inch of this epidermis of topsoil has been pored over and rooted at an infinity of times before; yet these hogs, in their everlastingly obsessive questings, go over each iota of terrain once and twice again. And God knows how much more before nightfall. Anyhow, I somehow feel a little better about them, and about being here with this crowd of porkers on such close, intimate terms. They seem to alter with exposure. Even the hog I first focused on in the stark light of dawn looks considerably improved—even vaguely respectable, in a way.

There's a wire fence separating this part of Hendricks' hog lot from . . . (I sense there to be a hog breathing on my back—G'wan, dammit, GIT!) from another section, equally large. The hogs over there appear to be a little bigger, but probably aren't. Some of them lie together in wads of three and four and nap. The fence that separates these two sections starts way up at the top of the slope near the barn and runs down to a stopping point at the edge of the pond, which means that those hogs on the other side of the fence have equal access to the pond and can come over here if they want to wade or swim. Otherwise, that other portion of the lot looks every bit as badly ravaged and ransacked: the entire com-bined expanse being composed of raw, woody desolation, seamed and trenched like a gutted, devastated battleground of the Great War, like no-man's-land. It's the Argonne Forest. And over there now, among the stumps and stones and jettisoned fragments of man-made waste, a lovely white-and-black gilt nestles herself, with sensual rolling wiggles, into a pillow of mud, and appears to drift to sleep.

With the further clearing of day, hogs and more hogs, hogs of all colors, a virtual aurora borealis of hogs, seem to arise and erupt from behind trees and everywhere. Their short, snorting grunts blend neatly with the couplings and conjoinings of railroad cars. There's a new string of cars over on the tracks, and every so often a diesel engine pulls up another, attaches it and huffs away.

Hogs are browsing on my hill. A young Yorkshire lifts his head, chewing, looks straight at me and offers, "*groonnk*," very tersely, like a haughty "hi." They are plowing up the entire topsoil layer of my hill, looking for miscellaneous chewables or whatever other things they may somehow have missed the thousands of times each one has been over it all before. They pick up rocks in their mouths, hold them a second or two, then let them drop. Whenever I make a sudden move the couple or three closest to me jerk away reflexively, but most of the rest (there are twenty-three in this part of the lot at the moment) stop and glower at me, each instantly striking a wideset frontlegged stance with the lowered head and the sort of disturbed, dead-eyed expression of so many rabid dogs. One, in disgust, runs up and bites the metal thing I'm sitting on. *Clang.*

Anyway, these rocks they uproot and . . . (good God, one of them has just chomped into a rock; no, I see now it's a hickory nut) . . . now I've forgotten what I was going to mention. Oh, these rocks have each been nosed or mouthed so many times that they're all worn almost smooth. There isn't a rough-edged stone around here. Even the few boulders look as if they've been rubbed slick by river rapids. Obviously, there's something in rocks hogs like. Maybe minerals or something. I know that they get iron from the dirt they ingest. In addition, of course, the body needs, if it is to maintain a perfect health, a certain quantity of phosphorus, lime, coal, sugar, gin, cement, rust, beans, mud and other bone-making elements. But these rocks! All so rounded! I surmise that two thirds—no, a mere *half* of the U.S. hog population would, if turned loose in Manhattan, have the buildings, megaliths and all stone projections crumbled and scoured down to manageable nubs within a matter of, say, eight to ten years.

A sweetfaced brown barrow has just scooched himself up a little mud nest right next to me and wiggled into it. He's lying here with his snoot pointed in my direction, munching on some new-found toothsome chewy and gazing hard at me, trying to figure my habits. Now his eyelids droop. He sees that all I'm doing is sitting cross-legged upon this metal heater grate, writing on a pad of yellow legal-sized foolscap. So he drowses off.

At 10:33 A.M. (C.S.T.), let it be recorded, an irate hog marched up from the right side of the hill and, apropos of nothing, flung a fresh-rooted dirt clod in my face and all over my notes. They root without end. Nothing discourages them, not even me. The Will to Root is the single most profound defining quality of hoglife. Only death will affect it—and then simply by postponing it until however long it takes to reach some great hoglot beyond. Still, there's such an admirable deliberateness about these hogs in this or whatever else they may happen to do. And, actually, the more I peruse them in their doings and comings and goings, the more logically shaped they seem, snout to tail. Somehow or other since I first arrived in the darkness of morning they've apparently reshaped themselves into bodily structures quite sensible and physically pure.

A train of flatcars stacked with earthmoving machinery heaves southward on another track about a half mile away. A raggedy Negro walking down the road stops and counts the distant cars. Idling hogs amble and squat. Some root. One sneezes. The sleeping hog beside me wags his ear a twitch or two and otherwise remains removed from the milieu. A Hampshire bites a Yorkshire's ear. A Poland China bites my foot. A white hog with a black face and black spot on his side executes a galloping gleeful leap into the vacant pond. A wandering rooter pussyfoots up the hill and sneezes right into the face of the one asleep, who responds merely with another quick ear-wag and continues his snooze.

Hogs are plopping down in various stances all around me now. Some sit propped upright on their forelegs, resting on their rumps, and face the road to watch for trucks or stare at the motionless freight cars. Some lie down. To execute this movement they first select a site, then kneel on their forelegs—a pose they sometimes hold for a few moments, making you (me) wonder if they're actually about to lie down or if they are engaged in the act of genuflecting—and finally lower their rears to the earth, all very deliberately and neatly.

There's an amplitude of birdchirps and crowcalls from the spindly branches. Now, too, a woodpecker. Over against a fencepost near the pond a hog scratches himself up and down, up and down, up, down, closing his eyes, ecstatic. Another hog has flopped next to the one who's been sleeping beside me. They face, touching, into the morning sun, and each occasionally makes small snuggling motions toward the other. Three beige-colored Yorkshires lap from the pond. Altogether, Ed currently supports about one hundred hogs around the farm and complains that it's getting hard to feed them what with corn having just gone up again for some reason.

Looming up on a crest of higher ground is a huge Gothic cathedral of a barn. Up here on the top of the slope the soil is strewn with a billion husked, stripped corn ears, all blended brownly into the mud (whereas all the land surrounding the hoglot is grassy-green). The barn sits on half a football field of level earth enclosed by a fence. The barn is used as a farrowing house for sows while the fenced space is a safe place where mothers can nurse their piglets till weaning day—at which time they graduate to the World War I battlefield pillage of the main hoglot.

From the rise on which the barn sits, the land falls away into a small, rolling valley that ends with the pond and roadside fence. I am on the highest sprout of earth within this valley, whereupon I have a splendid panoramic view of all these hogs as they perambulate among the trees or nap beside big rocks. The sun dies now behind a cloud. Hogs are slumbering in quiet clusters in every direction. It's nearing lunchtime—my lunchtime, anyway—but I'm going to stay glued right here to observe (and hopefully extract some Ultimate Truth from) this abundance of hogritude. I wonder if I have enough cigars.

PRELUDE TO THE AFTERNOON OF A HOG

Many of them seem to have accepted me now—or, in any case, feel comfortable enough to sleep within my arm's reach. And, too, they've ceased gnawing at my boots and the metal grate I'm sitting on. A russet Duroc rises from his dirt berth at the base of my hill, strolls to the pond to drink, then ambles back to his colleagues and drops back down among them with an emphatic sigh.

When I rise to answer the cravings beside a nearby tree, a full half-dozen hogs suddenly up and descend on my metal heater grate. They lift it with their noses together and fling it down the hill. Now four more join in to assault the grate, and I hear the brittle clang of their teeth on the steel. Even the two who'd been snuggled together asleep perk up and take part with the rest, all poking and sniffing and knocking at the heater grate. In incredible unison, they lift it above their heads and let it drop once, twice, again, again. It's hypnotic. Now they throw it on a pile of rocks and flip it over and over and kick at it with their agile trotters. A few saunter forth and poke me with their rubbery-tough snouts and look up to see if I'm angry over what the rest are doing. They've tossed, kicked or dragged my seat about fifty feet from where they found it and have now left it beside an uprooted tree. Most go back to their bedding sites and resettle. A smug black hog on the hill, a ringleader, scratches his ear with his right hind leg. He smiles.

I sit propped against a tree now, down a few yards from the top of my hill. The hogs have occupied the hill. Some of them are strutting possessively around on the crest or else lying back to bask in the near-noon sun. Some come down toward me and thrust up their heads—to gloat, I guess. A black gilt trots by and licks my arm. Then others drift over, vaguely apologetically, and let me scratch their snouts or tousle their ears—whereupon they shake their heads violently, like playful dogs, with ears flapping leathery, flapflapflapflappity. I oink; they oink back. A Hampshire has offered me a mudball lugged from the opposite side of the hill. Birds and hoggrunts and distantly droning sounds of heavy trucks all merge. Wind climbs, rolling over the hill and clattering against the dry, bare limbs of this tree. Two hogs meet at the foot of the hill, put their heads flat together like bookends, mutter something and flow off in separate directions. The rest poke and gruffle all around the lot, seeming to float just above the land like a flotilla of bristly dirigibles.

Time seems liquid. Five hogs are presently slounched down and snoring upon the topmost bump of the hill where I'd been sitting. A few birds alight from the trees to hop and dab among the hogs resting below. A light wind blows along the ground, and the yardful of recumbent hogritude rustles and shifts slightly. Skylight brightens and fades in irregular, lazy waves with clouds passing across the sun. The day has quieted. The pulse of hoglot life has slowed to a sensual languor, a drawn-out, low, slumberous grunt. It's mid-March on the winter edge of spring. It's noontime here in Tennessee. But back home in Georgia folks have been into their lunch for the past hour now, and the same thing holds for the people down in Ecuador. Yet in Rio de Janeiro, it's already four in the afternoon; and in the hot towns and snaky forests of Nigeria supper is mostly over, with the fiery daylight slipping rapidly away into dark night steam. And in the Crimea, on the Black Sea, it's 9 P.M. and nothing to do. It is midnight in New Delhi as well as throughout most of Tibet to the north; and in the Leyte Gulf, on the east edge of the Philippines, it has become 3 A.M., the really dark night of the soul. Nothing much happening there either. In Auckland, New Zealand, and out across the Tasman Sea the sun has just barely, *barely* come up; while along that horizontal band all the way from the Yukon down through San Francisco to Los Angeles and on southward, everyone is excited right now because it's 10 A.M. coffee-break time. And meanwhile, as all of this is taking place elsewhere on the planet, I've been rooted to this spot watching hogs sleep.

A hog in siesta on the hilltop has just jumped up to bump an intruding rooter down the slope, somersaulting to the bottom with a tumbling eruption of high-pitched squeals. Most of the hogs are up now, moseying about, perfectly unhurried: gambol and squat awhile, browse in the dried mud, drift in bulky serenity among the stumps and stubble and birds, call a sudden halt to it all every so often to look up at a sound or nudge another in the loin. Probe, poke, trot, root. Ah, hogs! They have unquenchably inquiring minds, each with a vast capacity for sustained wonder. And such a beatific quality—a certain handsomeness, really.

Comes a sudden whistle—now fading—and the spoke-and-hollow rumble from the nearby tracks rouses a few of these who were still at slumber. The others pause, gaze and then proceed with their solitary excavations. They are endlessly enthralled with things both around and beneath them and carry on with their daily lives mostly untroubled, except for appearing easily threatened by the abrupt motions of men. The tattered Negro who walked along the road this morning has come back from wherever it was he went. He sees me on this hill and waves, and a few pigs beside the pond look up at him, then turn to look at me as I wave back—thereby apparently resolving everything in their minds. Now they calmly lower their heads back down and go on about their browsings. I am not a threat. I don't have designs on their ribs or hams or all the sides of fresh, perambulating bacon they represent. I am pork with pad and pen, and I rest here upon this hill with these fellow hogs, not moving much and emanating, to the fullest measure possible, only the very *kindliest* of feelings toward my immediate surroundings.

I observe theirs to be a wonderfully savage marriage of animal and earth. Hogs share a perfectly accepted existential rapport with air and soil, operating at all times with a reverential respect for the body of the land. And, too, they carry out even their most humdrum functions with an almost aristocratic instinct for grace and form.

Up on the crest, near the barn, two blond hogs prance across the rocks, making playful squeals, faintly audible over the distance. Another, newer hog—this a blackfaced Hampshire with a strikingly clean, well-marked white saddle—has fallen asleep just above me, with his white forelegs stretched forth, his long snout couched neatly between his trotters, his eyes closed and the totality of his general bulk lying fully still but for the most *minute* flexings along the top edge of his metal-ringed nose, and a barely more frequent fluttering of his erected ears. He has notably long, handsome eyelashes. In his sleep, he sighs deeply and whiffs up a tiny roil of dust before his nostrils. A few yards away, another hog gives a gaping yawn and rolls from belly to side—still snuggling closely to a sleeping companion—then sits up on his forelegs, yawns again and waddles drowsily down to root up and chew a rock.

But it is of this first hog that I speak. He is, as I say, a Hampshire, slightly jowly, with white markings and a fine white tuft on his tailtip. I believe I saw him in the early darkness when I first climbed over the fence to get into this lot. He didn't run, as did some of the others, but stood his ground—though not in an angry manner—and then trailed me partway to the observation post I picked on the hill. He seems larger (being perhaps a month older) than most hogs here and conveys a certain *sang-froid,* a noblesse, which evidently derives from even the slightest degree of seniority in hog circles. He has richly brown eyes and a flair of remote wisdom about him, an air of assertiveness and inward strength of a sort that doesn't need to be demonstrated to prove it exists. Clearly, he is an eminent figure within the elaborate hierarchy of hogdom—at least among the hogs here. This is a coolly effective caste system which doesn't require a hog either to acquire or to hold his individual status by virtue of fighting. All he must do to make his position known in the hierarchal scheme is conduct himself a certain way. Even among peers of equal age and weight there seem to be signs and shared perceptions that determine the standing rank of Top Hog and underhog— all of which is maintained within a context of complete non-violence. An underhog would be loath to affront one of higher rank, and a Top Hog (such as this Hamp appears to be) would rarely need nor probably ever stoop to sully his tusks in combat with a subservient. Apparently, hogs seldom have serious fights.

A band of roving adolescents comes up from the opposite side of the hill to sniff and snort at me. When the boldest of them thrusts forward I reach out and stroke his snout. He jerks away, then slowly leans back for another touch; whereupon his companions, in succession, poke forth *their* snouts for a similar ritual caress, until, presumably satisfied, they pass on off over the hillside and into the trees. But this solemn, older-and-wiser Hampshire sleeps on. He rolled his head slightly to the side, cocked his right ear and opened his eyes for a second to examine the commotion, but quickly resumed his original pose, with a small letting-out of breath, and drowsed right on. Now a smaller Hampshire walks up, plonks down beside him for a second or two, then rises to nudge and lick at the hair on the older one's back. The sleeping Hamp flings up his head harshly and snorts a short, surly grunt, upon which the smaller hog ceases and desists.

The day is warm and nearly cloudless, wild with birds and, now and then, a squirrel. A redbird lights brightly on the Hamp's back, then takes off again to perch in a tree whose limbs overhang us both. The hog rolls, sleeping, to his left side. From across the fence, in the other section of the lot, there echoes a frenzy of whiny squeals and the sound of two crashing boars who bark, thrust and nip at each other in sudden combat—a brawl having to do, I guess, with some plush couch of dirt one of them had been warming. The redbird is frightened off by the noise, but the sleeping Hampshire doesn't stir. He is years and miles away from all this. He has extricated himself from the ruination of this war-scarred no-man's-land, this micro-universe of a hoglot, all grit and muck and ludicrous self-delusion, and is long gone off into the land of his mind, winging somewhere high up and far away, perhaps to wherever redbirds go.

His side heaves up and down a few quick beats and his right hind leg kicks back twice and then some more, like running. He's running. He's running with great effortless gliding bounds, unglued by gravitation entirely; running with uttermost fine abandon; running with centrifugal furious animal glee out across steep green dayshine and wide-arched air; running in vast grassy unfenced fields alive with tiny things of succulence wherever he might deign to stop and tarry. Other hogs are running, too—though none so swift as he—while still more brother swine sleep in the shadowcool of junglegrowth and leafy, shielding trees. The field careens along in dayglowing ripeness, swooping low, then up and over tender hills (beside each one of which lie just-right ponds of perfect cleanness, never touched before and full of fish). He gambols in open total sensual splendor, galloping across endless pastures, grass rolling underfoot forever; and everywhere he stops there spring forth soft neatly-made-up beds of fresh-stirred earth and pig-loving sun and sows, sows, sows, and nevermore disease or worms or parasites. He's a porcine prince, a poet, sage, a prophet, and he senses—deepdown throughout his stoutest girth—that he's at liberty at last to eat, to mate, to frolic and forage till his coathair grows long and gray. And, too, he's free from all those other upright creatures—the Man, Mister Charlie—whose kindly-seeming concern for his well-being is really just part of their act of sizing up his bulk for future certain slaughter and dismemberment. Here and for now he lives cut loose from past travails in fertile earthy splenditude, epicure eternal: bloomed awareness, eyewide, rising, risen, floated upon joy, fulfilled and foreverafter without pain. We leap and prance like wild things, bounding over lands where posts won't hold and hateful wire can never be nailed around us. We browse away long days in deepest viridescence, with snouts unringed to root in richest woods—as did the ancient ones—to root and root, root, root in purest exultation for tubers, acorns, grubs and supple truffles. Promise undefiled: the one and only pig among the planet's hundred millions who's guaranteed—hereby, henceforth—free passage through whatever may turn out to be the fullest-extended span, the uncut-short duration, of his natural animal life.

The Hamp grunts, eyes closed, and twirls his ears. An upturned clod from higher on the hill rolls toward him, hits, and he lifts his head. Then he hoists himself up to his forelegs, sits, blinks, yawns wide and smacks his lips, gazing dully across the brown, dirt-heaped hill and then on out across the swooping stretch of hoglot landscape with its rocky, tangled litter. Now he heaves himself indifferently down the hill, first to drink, and next to stroll slowly in cool pondmud rising high up on his trotters, almost to his shoulders.

When he emerges, he shakes himself dry, wags his tail, scratches his ear with a hind hoof and reclines along the ground once more—carefully laying down his body, but this time keeping his head erect to stare frozenly, and without sound, past the pond, fence, road and maybe, too, even beyond the tracks into the distant trees.

Things appear to take place in some oddly coordinated fashion here. First, for example, all the hogs will engross themselves in a sort of stationary rooting; then they may all spontaneously take to wandering in small groups to explore their turf; then sleep; then drink and prowl the muddy borders of the pond. Now it's scratching. Here are four acres' worth of hogs scratching their sides against trees, fenceposts, and every other available vertical. Those closest to me also seem even to wear pretty much the same silly smiling expression while they do this. It's as if the whole scene is orchestrated by something invisible. Then one will quit, followed by another and still others until they're all, as one, adequately scratched and ready to scour the terrain all over once more. Or fight. Hogs *will* fuss. Two small boars erupt in a minor skirmish until one of them scampers a short distance away and proceeds to sniff intently at a fallen treelimb, as if to show that the entire contretemps didn't really mean anything to *him* anyway, and how absurd it is for all of you (and me) to have been eyeballing something so inconsequential.

They are all quietly engaged in a great exalted dance. And now I am standing up on the very top of my hill with my arms outstretched, conducting them in their music, keeping the rhythm with one hand and gesturing toward those whose turns have come to perform. They don't seem at all self-conscious about this and continue to carry out their normal movements with perfect poise. Ahh.

Here and there, a bright young hog will up and sprint between the trees and—just as suddenly—settle down. Hogs move afoot in all manner of styles and cadences. The gait of a hog is a thrilling thing to behold. He doesn't merely lumber along like some dumb thing bound helplessly by fat, as the common public conception would have you think. No, indeed! For, from what I have observed, it is clear that a hog (depending on weather conditions and, of course, his mood) may: (1) *mosey*—this being an unhurried sort of just-walking-around, perhaps while in the act of considering something new to do; (2) *shuffle*—this is a halting, hoof-dragging pace usually employed while rooting, or sometimes while cutting up or clowning around with fellow hogs; (3) *stalk*—this is a stealthy, quiet means of converging upon something or other, which appears to consist of step-pause-lurk-advance; (4) *trot*—this is fun to watch, particularly at feeding time when a whole yardful of hogs abandon whatever they each may be up to and fall together into a stubby-legged run that makes their ears all flap up and down in rhythm; (5) *gallop*—hogs tend to do this one in small bunches; it's a standard horse-style run (allowing for differences in leg length), which is performed usually in an unobstructed field and executed with amazing grace, speed and *joie de vivre*. Now, there are probably lots of other forms of movement I've not yet seen, but the main point here is that a hog's gait is a varied and versatile thing—and that, when called upon, as the Germans say, "a hog is swift of foot."* In addition, by the way, hogs can also goose-step.

While I've been writing this, ten Yorkshires have gone wading. They're all over there just hoofing around in the water, gazing blandly at the surrounding scenery, or else immersing their snouts to snorf and poke at underwater roots and things—each one making sets of circular ripples that fan out across the pond and overlap with each other's. Sometimes they dunk their entire heads all the way up to their ears. Sometimes they duck just the tip of their snoots and blow bubbles. Sometimes they swim a stroke or two. Tradition holds that hogs can't swim because if they do, they cut their throats with their sharp front feet. Actually, they *can* swim. Wild hogs, in fact, have been found off the coast of Europe on islands they couldn't possibly have reached other than by navigating across at least twenty to twenty-five miles of open sea. And on top of that, a bevy of hogs *right at this very minute* is out there in front of me swimming along and damn well liking it. Or anyway they're not cutting their throats.

A young Yorkshire has wiggled under the fence somehow and is presently over in the strip of land between the road and railroad tracks walking around nosing things. There are now thirty-two hogs in this section of the lot, most of whom—aside from those in the pond—lie in scattered clusters, sunning. Clearly, hogs derive much of their energy and vibrancy from the sun. They are solar beings.

My friend, the older-and-wiser Hampshire, has been probing his way along the edge of the fence. He barks forth a brassy snort to drive away another hog who has tried to encroach upon his rootings. Now he comes

---

* "*Ein Schwein ist flink zu Füss.*" (I'm working entirely without notes here so you'll just have to bear with me, but I seem to recall that they put it this way. Or at least one of them put it this way to me one time. I have no idea why in the world they say this. I'm really not even sure they do.)

THE HOG BOOK

back up the hill whereupon he surveys me in fierce concentration for a full half minute with anxious, gentle, dark eyes, opens his mouth and works his jaws a bit, then turns and noses out a fresh dirt palate just a few yards away.

NOSES

Hog noses are fearfully and wonderfully made. (Not much is happening in the lot at the moment, so I feel free to note down a few observations.) Hogs appear to handle them much the same way elephants do, and, in fact, their snouts seem every bit as rubbery-pliable, proportionately speaking. The snouts are pink-tipped, leathery-fleshy truncated things, and usually wet—since hogs, like dogs, have no sweat glands and hence must perspire through their noses. The topmost rim is reinforced beneath the skin with bony gristle, which helps support the snout when used for tasks like overtossing rocks or shoveling up ramparts of dirt. Their nose-rooting functions are supported, too, by the very powerful bulge of muscle located at a point right behind their heads where their necks would be, if they had necks. As an old proverb puts it, "Nothing is more durable than a hog's snout." Hogs come on first with their noses. They lead with their noses. They thrust up their noses to you and advance until making a preliminary nose contact before anything else. They *know* through their noses—trust with them. Sometimes they even touch noses to talk with one another. I can't tell, though, whether they grunt through their noses or mouths or both; at least it does appear that they grunt certain grunts with their mouths closed. (The main occupational hazard in having a nose like this is *atrophic rhinitis,* a hog ailment amounting to a chronic head cold, which plays absolute hell with snouts, slowing the entire growth process of the hog and frequently causing untimely death. Sometimes a newborn piglet, in groping around for something to eat, will bump into his mother's snout, and if she has it, he gets it and then infects the rest of the litter.) Anyhow, insofar as for hog snouts in general, it somehow turns out that in spite of the incessant friction and use and abuse they are put to, the noses are always art-gum-eraser tender, sensual and moist—not all calloused or garmed up as it would seem they ought to be.

EARS

Then there's the matter of ears. They are very keen. Even when a hog's asleep his ears will perk up at noises and turn to one side or the other following moving sounds like radar tracking disks. A hog's ears may stand up erect or flop down toward the eyes somewhat like horse blinders. Aroused ears look like wide isosceles triangles lightly fringed with hair. Hogs have a great fondness for being fondled or scratched behind the ears and consider this a supreme act of friendship. On some farms, ears may have been notched by hogmen. Sometimes, too, they are notched by other hogs. Some people eat ears. Some others (fewer each year, it seems) persist in trying to make purses of silk from the ears of sows. And some people manage never to think about the subject at all. That amazes me.

TAILS

Tails are another thing. They (the tails of these hogs anyway) are, for the most part, not curled up at all times; rather they hang down like limp and very ordinary whipcords which wag every now and then, often for no apparent reason. The hogs here, however, *do* have fully intact tails —unlike in the pig parlors where, out of sheer boredom, young pigs acquire a penchant for biting off the tails of others. Some tails will coil themselves up in a semi-prehensile fashion, but generally they drape loosely across the rump like thin, stubby, disproportionate appendages which seem to be affixed to hogs only because hogs are animals and animals are just *supposed* to have tails, regardless of however perfunctory. When a hog is frightened or struck with some form of anxiety the tail will clamp tightly down across the rump and, as much as length allows, between the legs. Otherwise, it just sits there. You can look out upon an enclosed gaggle of hogs—as I am doing now—and suddenly everything turns into tails; all else simply goes out of focus. Or at other times it can be nothing but noses. Just so many disconnected noses walking around. Or ears: a billion ears. Seas of ears. Or maybe it's just me. Maybe I'm going insane. ("Laawd," said Ed, half laughing and shaking his head when I told him what it was I planned to do here, "*I* sho' as hell wouldn't wanta sit out dere all day with 'em.")

*   *   *

Forty or so hogs are now ringed around the base of my hill, rooting, snooting, bumping heads and moistly smacking their mouths. Soon they're going to root and grind this hill completely away in exploratory search of something I don't think they'll find. I wonder if they're the ones who excavated and toppled the trees that lie all around here. Everything is a potential meal. And every morsel of territory is imperative to the solitary pig upon it. Even this hoglot stands, for them, as a cherished thing, for hogs seem to need a notion of geographical place, a sensation of something shared—a sense of home.

It is four-thirty in the afternoon and I am the only non-hog who has been in the lot all day. Other than what the rest here have rummaged up for themselves, they've had nothing to eat, in the shape of a formal meal anyway, since breakfast. Breakfast was four hundred pounds of corn. Corn at 5 A.M. Now it's late in the day in the daily life of these special and particular hogs. And up in Iowa, at this moment, farmboys on tractors harrow up vast flat plots of soil to plant things for spring. While out across dust-dry empty wasteland reaches, distant trains plunge along and whistle thousands of miles, and maybe—as it must happen on rare once-and-again occasions—just maybe someone else out there hears. And meantime, far off into the ice-black North Atlantic, cumbersome cargo ships with skeleton crews lurch their way onward toward obscure ports, prowling through the cold, star-capped night in watery silence.

At four forty-five, a dozen hogs commence to eat a log. I'm glad to see it. After all these hours spent in penny-ante rooting it seems about time that they would finally realize they've just plain run out of places to look for the smaller sort of edibles. Here now is the damp, semi-erotical sound of combined Hog consuming wood communally.

They put rings in hogs' noses to keep them from rooting. I notice it hasn't slowed up any of this crew the slightest bit. Every single stone and solid object and atom of soil on this piece of acreage has been moved at least once. Or they've been studied or sniffed or pondered over or grunted to or urinated on or scratched at. To scratch his stomach, the normal hog will straddle a log crossways and undulate fore and aft. And sure enough, as soon as one does it all those around him follow suit. Everything is choreographed; it's all a cosmic waltz.

In the closing lateness of afternoon all the hogs are now off on a new round of rooting. They've moved away from the pond and gravitated toward the middle of the lot to grope and frisk among several overturned trees. A diesel switch engine towing a long load of trucks blares by, leaving in its wake the special desolation that always overwhelms a countryside when trains disappear. But not a single hog looks up this time. They are all heads down with tails aswish together in an insanely irregular tempo, tails which wag contempt at Fate. The birds of morning and midday have mostly quieted. The predominant sound now is a cacophonous overlay of singularly hoglike utterances, ranging from deep, gustatory grunts to a drawn out *baaawwwwp* whenever a strolling pig overextends himself into another's rooting room.

Hogs are extremely oral, of course, but there's more to it than just that. What with the usual circumstance of their living together, plus the fact that each one is an individually congenial creature to begin with, hogs have managed to develop not only a wide vocal range but a comprehensible system of verbal communication. It's obvious to anyone who's spent any time at all among them that—on the most elemental level—hogs employ squeals of varying pitch and volume to indicate fear, pain, hunger or sex urgings. Yet it goes much further. I suppose there must surely be some regional differences as well as peculiar breed dialects, but basic Hogolalia (the language of hogs) is a fairly structured arrangement of sounds, each, as I perceive them, with a distinct meaning. For the most part, the sounds of other animals are not so much a structured language as mere mechanical cries. With hogs it's different.

HOGOLALIA

To start with, there is the deeply pharyngeal, often guttural, *groonnk*, a more or less standard low grunt indicating that the hog is simply going on about his own daily business in a normal way. Then there's *rah* (sometimes heard as the classic "oink"—which is never actually enunciated), a quick, mildly snapped-off sound of acknowledgment in the spirit of "hello, how are you; what's happening," that sort of thing. *Wheenk* is a word usually repeated two or three times, amounting to a series of "what now?" oinks of minor irritation. *Rawrk* is the abrupt exclamation of startlement or surprise; while *Wheeeeeeeeiiiiii*—like a roomful of fingernails running lengthwise down a blackboard—is the sound of being hurtfully intruded upon. Hogs are able to explode in loud, barking, doglike *woof, woofs* (not to be confused with "arfs") to voice the sensation of outside danger. *Waaaaaagggh* is a warning to an adversary; and *Ooooooooeeeeeennnnkk*, which sounds more like a high glottoral sucking-in of air, or like wet hands wrenching a balloon, is a cry of shocked alarm and physical pain. More peaceful words include *howwwruff* or *hummphh*, the sound of pleasant discovery; *ronk*, a request for an adjoining hog to move over; and *baawwrp*, an oink of pleasure. Nursing sows often lullaby their young with soft, singsong *ruuuuuunnnnks,* and piglets return all manner of sweet light chatter into their mothers' ears. In addition to all these *conscious* sounds, hogs commonly snore in their sleep and sometimes even snort words, presumably in response to dream-visions.

Most of the sounds amount, in one sense, to small, vocal assertions of individual hoghood. But on top of that, they are mutually understood among groups of hogs as utterances useful for communicating facts as well as feelings; for a chief feature with hogs is that they are totally social, even communal, in their combined conduct. There are apparently no equivalents for swear words.

In certain situations (though I can't determine what governs them), hogs display a genuine sense of sharing—be it space, food, artifacts to poke at or almost anything else. They frequently scratch themselves upon one another and nose tenderly among the bristles of each other's backs, and appear, in general, to recognize that whatever it is they think they are, they are at least all together in being that thing. They're part of a seamless web of kinship. It's a transcendent communion, a joint partaking by each of this ecstatic awareness of being at one with one's race here, at home, upon common ground, where together they need never fear. They seem to take part in the mutual perception that they all suffer and enjoy in unison; that their lot in life is a special blend of elegance and despair. And, too, they sometimes even give hints of sharing a vague animal apprehension that they have each been brought to this place, or into the world itself, to serve out a short blissful prelude before some kind of common Apocalypse that will descend upon them somewhere aroundabout their sixth month of life.

So the main bond they have with each other, or with all living things in general, is the fact that they *are*. To hell with grandiose causes which project beyond a life span; the key chore is to root out some remotely sensible meaning to that which thrives inside a breathing body here and now. Hogs are classic existential phenomena. And because they sense this, they invest all their brief born days with those universally accessible traits that make their lives mutually comfortable: decency, candor, common sense, self-reliance, courage, integrity, love, charity and casual ecstasies. They are fully freehearted and accept among themselves the joy of simply *being*. Best of all, they are purely and totally what they are: not surrogates for anything else nor dumb animal objects to anthropomorphize. A happy hog carries acorns in his mouth.

❋ ❋ ❋

The day has grown golder and more cool. Up the slope toward the barn, on a flattened slab of rock, a fat white-and-orange cat sits motionless, squinting at the busied valley of hogs. The older-and-wiser Hampshire is lying beside a fallen tree. A smaller hog, a gilt, is propped against him with her head draped across his back, also sleeping.

Up the slope, the cat is gone. The birds are gone, too; and each of the hogs now seems to have separated from the others to conduct his or her functions in private, probing along at the close of daylight, most likely hoping for a last good, quick *root*. A totally black barrow struts up and sniffs at my shoulder and then into my ear. Hello! The countryside is silent: only a rare peep of nightbird, miscellaneous pig snufflings and the light sandpapery scrape of treebark wherever a solitary hog has stopped to scratch. One day is pretty much just like another.

My Hampshire friend is snoring in low sputters, taking deep breaths that shake the head of the gilt resting upon him. Maybe he's dreaming again. Maybe he's a horse. But maybe, too, right this minute as he sleeps here, his carcass meat is being calculated upon by a small-scale dogfood tycoon in Chicago. Or perhaps two months from today he will possibly appear in several scattered parts upon the plates of stodgy Rotary clubbers in Cleveland duly attacking their ritual pork chops (or some such) before hearing an inspirational speech by some ex-Kluxer turned stock broker.

"WOOOOOoooooooooo," up the slope. Up toward the barn and toolshed: "WOOOOOoooooeeeeeee." And hogs from all around raise their heads, and some begin to trot with brisk expectancy up the hill. "WOOOoo-uuupp." Hogs in the pond, hogs in the trees, and on the hill and in the rocks and now even my friend Hampshire—all are up and off and on their way to the crest of the slope, amid a din of high-pitched animal squeals, to where old Ed is shoveling grainy feed into the metal hopper that funnels into a feeder trough. I hear his half-drowned-out voice talking kindly to them, saying things I'm too far off to understand; yet I do sense the tone as one of great gentleness and a glad-to-see-them sort of feeling. And they, Lord knows, are glad to see Ed, too, and make properly joyful noises unto him.

Now, as they eat, they are hushed, except for chompings and smackings and the flappings of ears. And soon some of those who first arrived have had their fill and mosey back down the darkling slope. In time, there remains only a handful of hogs wolfing food around the bin, taking a few last, pre-bedtime bites. Others have gone back to their individual digs or else are just lying around, thinking. The older-and-wiser Hampshire fills himself to sufficience and drifts downhill to the pond to drink and await the onset of darkness. He strolls along the pond edge slowly and alone, a sovereign wanderer with a watchful and elegant mind, infused with profound dignity; but more than that, with an innate decorum, a regality. It's as dark now as it was when I first met him, but since that time he's undergone an awesome metamorphosis into a thing of beauty: He now appears to me to possess a wondrously elegant tuberosity of body with a compactness of design, all engineered to functional perfection, somewhat like a Porsche. He's become a whole new vision. I feel I've never really seen a hog until this moment. One of us has surely changed.

It is after eight in the evening. Elsewhere across America, balding husbands, stuffed with supper, are settling into their favorite chairs to gape blank-faced at thousands of bad TV shows. In New York, the theaters are filled. And in wild smokey plastic sterile cities throughout the land, nameless combos of union musicians are sawing away in the basement ballrooms of fifth-rate hotels, as smatterings of middle-aged couples sulk grimly through the tango. Meantime, some of the hogs here in middle Tennessee are already asleep. Others are solemnly grunting and snorting and nestling down for nightfall beneath a dark infinitude of ancestral stars, brotherly stars.

The Hampshire has hollowed out a saucer of mud and squnched himself inside it, being ever so careful not to smudge his fine black-and-white coat. He lowers his head. His eyes blink heavy-lidded. And in moments he's tapered off into a soft, untroubled slumber. His sides move in and out with long, swelling breaths as he melds away into fields replete with perfectly-formed flowers and scenes of great peace, mildly heroic exploits, modest hopes fulfilled . . . probably most of the same simple visions he's glimpsed in his sleeping mind so many, many times before.

Good night, sweet hog! The rest is silence.

VI
Pig Poetry

Historically, the hog is the focus of some of civilization's finest out-pourings of eloquent poetical reflection. It is clearly impossible to include here even a significant percentage of the vast and wide-ranging array of verse published across the years. For historical perspective, the opening poem in this distinguished selection was written in England in the eighteenth century. The rest are lines and lyrics inspired by simple contemplation upon the porcine condition by several contemporary poets as well as proponents of the Hog Movement.

ODE

To a Pig while His Nose was being bored.

Hark! hark! that Pig—that Pig! the hideous note,
　　More loud, more dissonant, each moment grows—
Would one not think the knife was in his throat?
　　And yet they are only boring through his nose.

Pig! 'tis your master's pleasure—then be still,
　　And hold your nose to let the iron through!
Dare you resist your lawful Sovereign's will?
　　Rebellious Swine! you know not what you do.

To man o'er beast the power was given;
　　Pig, hear the truth, and never murmur more!
Would you rebel against the will of Heaven?
　　You impious beast, be still, and let them bore!

The social Pig resigns his natural rights
    When first with man he covenants to live;
He barters them for safer stye delights,
    For grains and wash, which man alone can give.

Sure is provision on the social plan,
    Secure the comforts that to each belong!
Oh, happy Swine! the impartial sway of man
    Alike protects the weak Pig and the strong.

And you resist! you struggle now because
    Your master has thought fit to bore your nose!
You grunt in flat rebellion to the laws
    Society finds needful to impose!

Go to the forest, Piggy, and deplore
    The miserable lot of savage Swine!
See how young Pigs fly from the great Boar,
    And see how coarse and scantily they dine!

Behold their hourly danger, when who will
    May hunt or snare or seize them for his food!
Oh, happy Pig! whom none presumes to kill
    Till your protecting master thinks it good!

And when, at last, the closing hour of life
    Arrives (for Pigs must die as well as Man),
When in your throat you feel the long sharp knife,
    And the blood trickles to the pudding pan;

And when, at last, the death wound yawning wide,
    Fainter and fainter grows the expiring cry,
Is there no grateful joy, no loyal pride,
    To think that for your master's good you die?
                                    —Robert Southey

HOG HAIKU

Fall: with soulful grace
    the heavy-hammed swine sinks
down in ooze, and ahs.
                    —Frank Trippett

PIG SONG

Si yo creyera que mi repuesta sería
a persona que pudiera hablar con el agricultor,
este cerdito nunca más charlaría;
pero, porque tu no puedes convencerle de
que un cerdo ha hablado contigo,
voy a decirte unas cosas en confianza.
                          —Juan Valdez de Santa Toledo

Let us go then, you and I,
where the field yields to the sty
like a "See Rock City" poster on a stable;
let us go through certain half-erected roosts
where some ducklings or a goose
speaks of sleepless nights in one-perch cheep chicken pens
and barnyards swept with sins;
roosts that reek with stale fowl smell
and birds pell mell
that make you want to yell some swelling question . . .
come on, now, ask me, "What is it?"
No? . . . well, let's take a walk.

In the sty they come and go
speaking of the rodeo.

The moist hay odor drifts along the gutters,
the green smell that steps up and mutters,
licks my back in the corners where it itches,
lingers among the trough and sputters,
passes through the farmer's kitchen who,
thinking his wife fast asleep at last,
reads lewd magazines to escape her bitching.

And indeed there will be time
to tell you of the smell that glides along the grass
and drifts along the gutters;
There will be time, there will be time
to make a squeak to greet the squealers I will meet;
and time for all the playful things
with all the luscious treats to swallow;
time for you and time for me,
and time yet for a hundred incisions
and for a hundred divisions
and being served with toast and tea.

In the sty they come and go
speaking of the rodeo.

And indeed there will be time
to blurt out, "Eat a peach!" or "Eat a pear!"
Time to slide into the mud kersplat
like falling in a jello vat.
(The roosters will say, "My, his legs are getting fat!")

My double chin, my rounding rump
my shoulders bulging in a clump
(The ducks will mutter, "But how his belly is getting plump!")
Do I dare
Get up from the mud?
In a minute there is time
for incisions and divisions to make a ham of stud.

No! I am not a fierce wild boar, nor was meant to be;
am an old stud hog, one that will do
to swell a sow, start a litter or two,
no doubt with an easy tool,
deferential, glad to be of use,
chubby, stout, ridiculous,
full of pounds, but a bit obtuse;
at times, indeed, almost obscene—
almost, at times, a piggy bank.

I grow round . . . I grow round . . .
I shall be measured pound for pound.

Shall I venture to the trough, do I dare to stuff with starches?
I shall dine on low-cal tubers and go on diet marches.
I have heard the farmers talking each to each.
I hope they will not come for me.
I have seen them riding their large white mares,
combing the white-haired mane for the fair
where the judges weigh and then compare.

We have lingered in the corners of the pen,
amid the mire, slime, and swill strewn through the sty
Till human voices "So o o o ey!" and we die.

—John Southall Hatcher

I SEE BY YOUR OUTFIT
THAT YOU ARE A DUROC

A porker came in from the highway this morning,
  All battered and bruised with his leg in a sling.
He cried, "Friend, please heed me, my hours are numbered,
  So hear my sad tale and the tidings I bring."

"I see by your outfit that you are a Duroc,"
  I said to the swine as he came into sight.
"Come forward, good comrade, and rest by the creekbank,
  Unlimber your burdens and tell me your plight."

The hog stumbled forward and sank to his haunches;
  His cheeks seemed to tremble, his hooves were blood-red.
"Beware the big trucks," he panted and whispered,
  "For where they are going is laden with dread.

"Ah, 'twas once in the hoglot I wallowed and frolicked;
  The randiest boar that you ever did see.
I ate all my corn and did what was expected,
  And never had reason to think I should flee.

"My tail curled and wagged in a come-hither manner;
  My ears stood erect and the sows would all sigh.
The barrows and gilts and all of my trough-mates
  Acknowledged my standing as 'Prince of the Sty.'

"I could foretell a rainstorm, see visions in dirt-clods;
  I could think and could sing and could even do sums.
But one autumn day in the bloom of my fatness
  I learned the cold fate to which each hog succumbs."

He paused and he coughed and looked up at the heavens;
  Only then did I see the deep hole in his throat.
He fell to one knee and his eyes became foggy
  As blood trickled freely and stained his fine coat.

"Oh go to the hoglots all over the landscape,
  Oh summon all swine and make sure that they hear:
I have come from the place where all porkers are taken,
  And once they arrive, nevermore reappear!"

He fell to one side and his breathing grew fainter;
  I knelt and I cradled his head in my arm.
"The men we have trusted are waiting to slay us;
  Please try to assure them we've meant them no harm."

Small beasts of the woodlands assembled beside him;
  The birds and the butterflies perched on his breast.
They lifted his body and carried him softly
  Deep into the forest and laid him to rest.
                              —William Hedgepeth

QUO VADIS PORCUS?

(Little Hog Lost)

The cow is standing on the porch,
  The goat is on the rooftop;
Old mule is trudging up the drive,
  I hear his sodden hoofclop;
The sheep sit round the table
  Where I bid them all to dine;
But dark grows near, the hour's late
  And where oh where's my swine?

I asked him come 'bout half past six
  To share our fine repast;
I hope he'll make his entrance soon,
  Our food supply won't last.
The horse is here and so's the bull,
  The geese and chickens too.
I only wish my hog would come
  Before we all are through.

We're seated in the dining room
  Enjoying a buffet
Of bacon, ham and sausage that
  The cook prepared today;
And chitterlings and spare ribs
  Plus pork chops, spam and brain.
I can't imagine why my hog
  Has chosen to refrain

From showing up to sup with us
  And share a jolly time.
He'd add so much if he were here;
  Oh where oh where's my swine?
                              —William Hedgepeth

PIGS' EARS

One Saturday night
to make an event
I bought pigs' ears
at the Colonial Store.
It would be something
to talk about I knew
as I asked the Negro woman
at the meat counter what
I should do with them.
Boil them for an hour.
Which made the leathery
cartilage pink and soft
as any young thing's ear
on my cannibal plate like
a last-minute blind date
(and deaf).
           I really had
rather eat a silk purse.
                    —Coleman Barks

WAR MEMORIAL

In a pine thicket, honeysuckle and briars
The Hog Pen rusts in the sunshine.
Once six stout-hearted volunteers (as much
volunteers as we all are) waited:
Amid yesterday's ration of scraps and
bacon grease and Merita to give their
last full measure in a war I read about
in books.

Later, in a war I read about in directives,
I opened an olive drab can of 1945 hamloaf.
And remembered.
                    —Ralph McGill, Jr.

THE VIEW FROM A DRIVE-IN
NEAR SANFORD STADIUM,
AUGUST 1974

The smog
ambled in on little hog's feet
snuffled over the neon dancing pig
down through cinderblock niggertown.
Roll, tide, roll.

—Ralph McGill, Jr.

NATURAL HOG LIMERICKS

A testy old pigherd named Stein,
When irked by his hogs would yell "Swine!"
    "Well look," snapped one pork,
    "If you want to work
With bluebirds, try some other line."

You can't put a hog in a zoo—
There's no telling what he will do.
    He may get loud
    And charge at the crowd
Or just lie there not looking at you.

You can't get a hog too worked up—
He's liable to jump up and whup
    Your horses and you
    And anyone who
Says anything to him but "Yup."

You can't let a hog get run down—
The worst he'll do to you is frown,
    But then too the best
    He'll do for you is rest,
Without even* gaining a pown.

—Roy Blount, Jr.

*If he can help it.

SUNHOG

Volatile and vital,
baking brittle all it
brushes, beams of heatful
sun sheer down through
shrouds of overhanging
leafless limbs and peer with
splintered light upon a
sleeping hog.

Sun leans hard with searing
beams of fire that work their
way into the soil and
vainly seek to coax some
show of life and growth from
picked and desolated
pigsty mud.

Nestled sweetly in a
gout of dirt uprooted
for a resting site, the
slumb'ring swine strolls cool and
proudly in some purely
private landscape looming
softly deep behind his
tight-shut eyes.

Far past reach of weather,
deep adrift in balmy
hogdream, he can swirl in
browns and greens and all the
lushly porcine scenes whose
secret glory far out-
shines the fertile fiery
sun itself.

—Possum Valdez

PIG THOUGHTS AT NOON

a vegetarian stroked at noon behind my ears
mumbled about my being bred for death (his pun),
but his thoughts were elsewhere

among feathers, furs,
rare flaming symmetry
and outspread wings, not me;

for though my soul dwells beyond swift stallions
and above the tree-couched cat
I am groomed for termination
and no one mourns my passing.

He means well, I suppose—
my friend the vegetarian—
but when he tries to find comparisons for me
his mind wanders to the puma, the cheetah, the jaguar
(the sum of whose lifetime thoughts
I could formulate on one hoof);

socratically he tries to penetrate
my crude surfaces
but is stopped by the shadows of things:

he cannot caress my short hair, bulk, snout,
cannot remove himself from reflexive imagery—
the stuck pig still squeals
no pearls are cast before me
I am symbol for greed and
things remote from godliness
forbidden as vile to some
but devoured at every part
feet, brains, joints, entrails.

So it is that I
become each of you
and am your metaphor—
what you seek as you peek behind the surfaces,
for who has sensed the nobility in my pig heart
and has caught the glint in my eyes
can ponder the beginning of the universe
and probe the heart of man.

—John Southall Hatcher

THE LAMENT OF SOW 55

You lured me with the juice
Of swinging jowl, on spear
Of shining tush impaled me thus,
In meadow thorn and bristled blush
Taught me wedded death.

Now I wait
Amid the fetid droppings
Of those who go before,
Feel their dying heat fan memory
In harsh caress.
I hear again the nasaled love-song's dirge,
Lament I took to heart
And wept upon,
Believing then your destiny ill-starred;
I saw with rounded eye
Your noble neck pricked crimson
In the crash of iron gate.

So I opened there,
Yorkshire pure in greenest love,
Pressed to heart the tusked and bristled thrusts
Of painful pleasure puncturing the soul
Til redness filled gray skies
And splashed the clouds sienna.
Afterwards, you left,
Never knew nor saw—
Your eye reached others where
Even before the red-rimmed sun, unlidded,
Peered into day;
I saw you, even then, unspent
Midst meadowed moonlight
Stalking other virgin sport
In grunted echoes.

The mutterings grow soft now,
Before the silver dawn
And wraith-gray gates that shiver,
Yawn, stretch up and fling awake,
Embracing destiny,

As I, crouched in blackness,
Feel the ancient wisdoms catch my throat
While far above, the crossing stars
Pass by your shadow blithely
Whispering my name.

      —Bonnie Sheldon

THE HOG POET

Everybody lie back
& have a snort.

Spread the goodies out
where the sun can hit.

porkchop, fatback
hackles & hooves.

Asspay the aconbay
Asspay the knife.

Food is the before.
Song the afterlife.

    ❋ ❋ ❋

Compadres, ah
cement compadres,

this is not all
there is—trough,

pool & nap. Gruffle
through your nose

that gods will wake
& feed their children

counting piggies
in their toes.

     —Coleman Barks

THE MIGHTY ANGLO-SAXON HOG UPRISING

Yorkshire, Hampshire,   Berkshire and Duroc,
pigs all afile,   brave boars together.
Battle-hard Sidney,   their swinely swift liege pig
to his war-porkers spoke,   to his stout horde began:
    "Famed shield-hogs,   offspring of ealdors,
fierce battle-times   in yore-days many
have we abided   the ringing of hand-swords,
the crashing of heirlooms,   adorned battle-blades,
now to us is come   our renowned victory-time.
Endure it we must   the shedding of boar blood
for kinsmen, our brothers,   alone in the ground;
better to avenge   one warrior well
than sadly to mourn   ten thousand a lifetime
without battle-sweat."
    Thus from afar   could their loathed-one see
these pig-troops assembled,   the war-ready bands.
From barrows and burgs   they watched hog-hordes heave high
bright banners and ash-spears,   gold shields and bold blades
and stout battle-weeds;   they heard grunting and squealing,
with reason they trembled,   with swine-fear they dreaded
the resolute boar lords,   the hog-rush to come.
    Then as I have heard   in the gabled sty-hall
spoke the famed of the battle-shoats,   the high-minded Lothar
the choicest of trough-lads: "Hear now my mind-thoughts,
valor-clad war-hogs,   for the gem of the heavens
now nears the hilltops;   not longer may we
abide here in peace.   Sidney our liege lord
has need of our brave deeds,   glory-work in warfare
with the two-footed foe.   He it was who gave us,
the high-sitting pork lord,   these morsels for munching
and burnished gold nose rings.   Now has the time come
when our corn-giver good   has need of our swine strength,
kin-pigs together.   Now will we barter
loaned-life for glory,   win fame for the far-herds
and acorns for rooting."
    Then were the swift ones
warlike to see there   eager for battle.
Straight-way the terror   was made known to the pork-foes,
the wretched ones,   as I have heard;
throughout the middle-yard   was told to all

of proud-marching boar-bands,   the war-ready swine,
of Chester White, Landrace,   Poland China and Hereford,
of Large Black and Tamworth,   breeders and porkers,
all shining in war-gear.   Not as in yore-days
when hog-hearts were humbled,   but stepping with snouts high
they went westward and eastward,   amid forests and walled-burgs,
near nesses and high-halls   and farm dwellings many.
With vengeance and hot ire   they broke loose the pen-bonds,
marched through the wide plains   and righted old wrongs.
    Those were good pigs   who quick comfort gave
to Sidney their leader,   the wise lord of oinkers,
and Lothar the young shoat   who good counsel gave.
Grim were the trough-friends,   scathers to keepers,
hard to the hog-foes   who fled from the land.

                     —John Southall Hatcher

(NOTE: the poem designated as "The Mighty Anglo-Saxon Hog Uprising" is actually a titleless manuscript whose history is as interesting as the poem itself. The manuscript was found in York in the late part of the sixteenth century by a wine merchant who brought it to the attention of the authorities. One of the few Old English poems to be left out of the Exeter Book, Vercelli Book and Junius Manuscript, this poem probably dates back to the early seventh century when these events actually occurred. According to the merchant who discovered the work [a John de Hamtoun], there was originally a note attached to the poem which read: "If my readers cannot believe what I herein depict in the tradition of my fathers before me, I can only say that I actually saw the massed pigs trampling over the land taking control of the country without regard to the sentiments or intentions of their victims, and so what I write here is but a portion of the story that could be told, and that someday must be told if there is to be any hope whatsoever." I have translated and edited this poem as best I could to capture the richness of the Anglo-Saxon imagery and the emotional impact of the poem. I only hope the message of the work is as clear as the lines are beauteous.—John S. Hatcher)

HOG HEAVEN

CHORUS: Hog hea-ven hog hea-ven just one step a-bove

Moun-tains for think-ing and val-leys for love

Pas-tures of plea-sure and a-cres of mud

VERSE: From the low lands of Geor-gia to the high sea-coast shore, my

broth-ers have been hound-ed for centuries or more.

And it's high time for chang-ing and set-tling the score.

Don't want my place on the food chain no more.

(CHORUS)

2. Leonardo daVinci's got nothing on me.
Beauty's in the eye that beholds it, you see.
Einstein and Shakespeare had nothin' but dreams,
And I'm entitled to my share, it seems
(CHORUS)

3. Farmers are helpless, butchers are blind,
Merchants are ruthless and people unkind.
We danced in the marketplace, we all stood in line
But we're headed for places that are better this time.
(CHORUS)

4. My days are spent watching, my time is unsure;
With each swing of the blade, my Karma's secured.
With each passin' century my numbers increase
Let us pray for the species that takes pride in the feast.

Words and music by Michael James Catalano
©1976

VII
The Porcine Potential

"Then," said he, "the pigs are a race unjustly calumniated. Pig has, it seems, not been wanting to man, but man to pig. We do not allow him time for his education . . ."
—from James Boswell's *Life of Johnson*

For, you see, this is not an easy time to be a hog. Certainly not for the moment, at least. Because under the existing reign of things, our society as a whole currently holds hogness in public disfavor. The hog, for the most part, is officially regarded as some sort of mindless porkheap hopelessly bogged down in a mire of distasteful personal habits. Few other creatures can measure up (or down) to him in terms of popular disrepute.

Therefore, rather than dawdling a moment longer we must plunge straightaway to the essence of the issue, which is this: Given the way things are, what is a well-meaning hog to do? The problem, you understand, is compounded by the fact that the vast majority of hogs are cut off from social intercourse, confined in places out of the mainstreams of our consciousness, vilified and rendered altogether helpless when it comes to defending their honor, much less their lives. In the meantime, the custodians of respectable society and conventional wisdom bask in their own delicious sensations of abhorrence and loathing—while simultaneously seizing upon every possibility for perpetuating what amounts to an almost completely anti-porcine *status quo*. Hogs are a mirror of people's bad consciences.

Snout-ringing, for instance, is one of the more obvious symbols of such repression. In days of yore when hogs roamed more or less free, most hogmen found it necessary to clamp painful rings on the rims of hogs' snouts to prevent, or at least discourage, them from rooting up crops. Yet even today, with most hogs enclosed in the barren ravagement and bucolic purgatory of typical hoglots, the practice persists—though now it ranks as more of a simple, sinister (though probably, in many instances, subconscious) attempt to thwart hogs and keep them in their place: in a lowly sort of social lock-step, so to speak. Hogs definitely do not live high on the hog.

Yet people continue to come to the hoglot, and they come for many reasons: to play, to plunder, to worship, to learn. And some of those who learn come away with genuine insights. "Many extraordinary examples of the docility and intelligence of the too much despised hog are on record," declares the author of a 120-year-old text, "however, except in a few isolated instances the hog's education is utterly neglected; all it has to do is eat and sleep, and become fat—its utility to man commencing at its death, by the knife of the butcher."[1]

Indeed, the system's universality and its sheer arbitrariness are such that hogs are nowhere permitted the chance to exist and evolve, and possibly thereby to enchance all the rest of civilization in some entirely new capacity. They are held instead in perpetual abeyance. Even a set of newborn piglets, freshly gusting with vitality and spirit, is relegated to the status of just so much squelched flesh—flesh forever hinting briefly at lives we will never see. For from the very instant of delivery each litter brings with it into the world a sense of great adventure missed.

The essential struggle here is over the fundamental question of the hog's future identity. This is an awesome barrier to overcome inasmuch as it flies in the face of all conventional respectabilities. As it stands today, insofar as most of Western Civilization is concerned, even the most rudimentary present-day skills of hogs—not to speak of the great untapped porcine potential—have remained as unsuspected and lost as the continent of Atlantis.

"In general, there is nothing in the life of a hog, in his domesticated state at least, which calls for any exercise of reasoning powers . . . All his wants are anticipated, and his world is limited to the precincts of his sty . . . Yet even in this state of luxurious ease, individuals have shown extraordinary intelligence."[2] This is a key paradox of hogritude. But it is one of which experienced, dedicated hogmen have long been aware, torn as they are between commercial and metaphysical considerations. Hogs are consistently rated as exceedingly bright on the basis of barnyard observations by farmers who—even though most have neither the time nor disposition to make a formal study of swine sagacity—just can't help but notice such things and marvel quietly at them.

But not just farmers. Others who find themselves in the presence of hogs for any period of time usually come away with their own glowing pronouncements on pig intelligence. Like Abraham Lincoln, who observed, "A pig won't believe anything he can't see." And a staff writer for the ultra-urbane *The New Yorker*, who perceived and recorded in an article that "Pigs are not only smart and tidy but sensitive, affectionate, playful and brave." And Representative Schwengel of Iowa, who declared, according to the *Congressional Record*, that "the wonderful pig [is] the most intelligent domestic animal in America, according to a study made by Cornell University."

A more official scientific commentary on hog I.Q. resulted from recent tests judging the comparative intelligence of various animals as conducted at Bar Harbor, Maine. This was a maze-type test, given to animals that had never before been exposed to a maze, to see how long it would take them to figure out the correct path. While horses, dogs and even chickens could be trained to navigate the maze (cats refused to participate), the pigs came out on top because of what was said to be their capacity for "independent thinking and ability to figure out a problem for themselves."

The Bar Harbor tests confirm similar laboratory studies elsewhere which, in turn, simply serve to corroborate informal findings back on the grass-roots level among rural folk who witness firsthand the superior brain power of pigs in action—and, as a result, are almost invariably respectful. "They's just no *tellin'* what all a pig can do," vowed a Virginia hogman, expressing himself with more reckless exuberance than one normally hears from farmers. In general, hogmen are outwardly taciturn and sober, as is their wont. Yet in their innermost heart of hearts and deep lost caves of their nights they are secret hogfreaks. Even though your typical hogman tends to play things close to the chest, many will eventually confess that they actually get high off hogs. In numerous cases across the country hogmen have been discovered to possess framed photos or paintings of hogs (icons in some instances) hanging hidden deep inside their closets or private places. These are the men who know what it's all about around the barnyard and who therefore cannot help but carry on with their daily lives in a constant state of hog-awe. As George Orwell wrote in *Animal Farm*, ". . . the work of teaching and organizing the others fell naturally upon the pigs, who were generally recognized as being the cleverest of animals." And later, of course, it was the ruling clique of pigs who scrapped all the hitherto sacred commandments of the farm in favor of the one that stated: ALL ANIMALS ARE EQUAL BUT SOME ARE MORE EQUAL THAN OTHERS.

In fact, in a number of measurable ways hogs are indeed more than the equal of others. For instance, hogs have a proportionately larger brain than oxen, sheep, cows and many other farm animals; and, too, they have a greater preponderance of that part of the brain that governs reasoning ability. "There are anecdotes enough to prove them possessed of memory, attachment, and social qualities; but at present the system of treatment affords no scope for the development of any but mere brute instincts."[3]

Says Dr. Marian Breland, co-founder of Animal Behavior Enterprises in Hot Springs, Arkansas, "An animal is smart in a way that can be measured: If it learns quickly what will reward it and what will not, and if it can make quick deductions from minor clues—that is, recognize the significant difference between objects that resemble each other. This is certainly true of pigs. They will generalize from one situation to another very readily, so that if you put them in a new situation where things are a little bit different they'll go ahead and go to work anyway, whereas a more limited animal will act lost."

Marian and her late husband Keller Breland, both graduate psychologists, became involved in 1942 in a government research project that has only recently been declassified. This project, which involved the conditioning of pigeons to guide missiles, was a direct application of what she calls "new animal psychology" to the control of animal behavior. In 1947 they set up Animal Behavior Enterprises, with its training center located now on 260 acres outside Hot Springs, where thirty technicians work away on animal research as well as on the production of special animal shows that appear all around the country. Cockatoos ride around on bicycles, ducks plunge from diving boards, raccoons play basketball, groundhogs raise the Confederate flag, roosters walk tightropes—that sort of thing. A.B.E. turns out several hundred animal graduates a year, in-

Hog Butchery Chart

THE HOG BOOK

cluding otters, crows, coatimundis, sea lions, killer whales, monkeys, foxes, rabbits, rats, cats and dogs. On the basis of her definition of intelligence, Dr. Breland rates certain primates first, followed by (guess who!) pigs . . . followed next by porpoises. (This is interesting because porpoises were at one time, and are still in some places, referred to as "sea hogs.") She places dogs below porpoises, then cats, crows, parrots and so on down the list.

"Now dogs," she says, "have traditionally been bred for specific things like birding, pointing, flushing, retrieving, things like this. We know now that hogs *can* be trained to do all of these. I don't recall an instance of a hog being used for herding, but I think it could be done. We know they've been used for retrieving, finding lost articles and so forth. Yes, I would say"—she reflects after a slight pause—"that if anybody took an alert hog and really worked with it, that hog would learn everything a smart dog could learn and maybe lots of additional things—just on the basis of their generalizing ability." She points out, however, that one has to take certain physical limitations into account, such as the fact that pigs have poorer eyesight than dogs, but that otherwise "pigs condition very rapidly and are among the most tractable animals we've worked with." Still, she shrugs, "a few farmers think pigs are very stupid, probably because they've tried to control pigs by force. Well, you just can't do this. They're very poor animals to try to push around. They're big and they're stubborn, so if you try to force them to do something your chances of success are quite poor. But if you treat them with respect they learn very quickly. They're very responsive."

What keeps coming up in all these explorations into the deep infernal caverns of the hog mind is that hogs are indeed intelligent, but more important, they possess definite character, integrity. "I have a friendly feel-

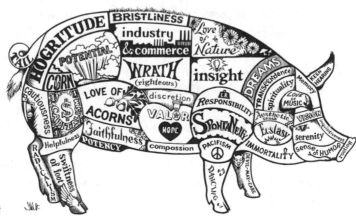

Hog Phrenology Chart

ing toward pigs generally and consider them the most intelligent of beasts, not excepting the elephant and the anthropoid ape," wrote naturalist and novelist W. H. Hudson, author of *Green Mansions*. "I also like his disposition and attitude toward all other creatures, especially man. He is not suspicious, or shrinkingly submissive, like horses, cattle and sheep; nor an impudent devil-may-care like the goat; nor hostile like the goose; nor condescending like the cat; nor a flattering parasite like the dog. He views us from a totally different, a sort of democratic standpoint as fellow citizens and brothers, and takes it for granted, or grunted, that we understand his language, and without servility or insolence he has a natural, pleasant *camarados*-all or hail-fellow-well-met air with us."

Clearly, then, hogs are creatures of boundless charm and enchantment, largely because they are clear-headed, perspicacious beings with feelings. It is this aspect that opens up to us the entire emerging field of hogology, the study of the mind of the hog. For with all things taken into account, hogs are better than just smart: They possess definite senses of self-awareness, discernible personality and a capacity for creative mental growth. Speaking on the floor of the U. S. House of Representatives, Congressman Scherle of Iowa displayed a touching color photograph of a hog and observed, "Hogs are not ony beautiful, they are perhaps the most intelligent of animals—sometimes perhaps exceeding man himself. Each is an individual, with its own mind, and does not willingly follow the herd blindly. The hog has little fear and rarely is found to abandon its responsibility for its offspring. It knows how to relax, but also knows how to work, as it constantly seeks freedom . . ."

These characteristics are eminently visible in single hogs, but they emerge even more profoundly in the distinct intimacy of hoglot life, wherein their group existences consist of collective acts of individualized decency and a sense of grace under pressure. Now, just to keep things properly in perspective here, it is not that there is no fault whatever to be found with pigs. No, indeed. Some have been found to be rude and not especially attractive. Some are clearly unsophisticated and not notably alert. Some are lethargic, as if depressed or bestruck with some moldering *Angst*. On the other hand, however, even amateur hog-watchers throughout the land are daily coming upon scores of breakthroughs in hogology that more than offset whatever personal failings hogs may be said to have at present. There is, for example, very minimal competition among them on a day-to-day basis. Moreover, they are by no means fawning, dependent creatures, but are self-sufficient, resourceful, somewhat insubordinate and at times even a bit saucy, a bit given to idle tomfoolery.

Whatever minor faults some of them may commit are certainly excusable in the final analysis; for if what some say is true, hogs each come into the world with vague premonitions of doom already built into their being. Perhaps it's as a result of this that many people have noted the sense of fey humor that comes forth in their daily carryings-on, a somewhat existential attitude of jocularity. As claimed one young Iowan, "Hogs know they're gonna be killed so they get this I-don't-give-a-shit attitude; it's like they're aloof and amused at all the crap going on around them." They are deliberately existence-oriented. In a poem on the pig written in 1799, Robert Southey said:

> ". . . The last charge—he lives
> A dirty life. Here I could shelter him
> With noble and right-reverend precedents,
> And show by sanction of authority
> That 'tis a very honorable thing
> To thrive by dirty ways. But let me rest
> On better ground the unanswerable defence.
> The pig is a philosopher, who knows
> No prejudice. Dirt? Jacob, what is dirt?"

It is not uncommon for people nowadays to be genuinely concerned with the personal problems and mental happiness of those animals they are most intimately involved with. And this is as it should be, because in the most ultimate sense you do become responsible forever to what you have tamed. Animal psychiatrists, in fact, generally counsel pet and owner together on the assumption that they share psychological problems in the same sense that they share domiciles. (A pet cemetery in the urban Northeast carries this notion to the extent of guaranteeing that pets owned by Negroes aren't acceptable.) But for the most part, neither owners nor professional animal counselors have made much headway (most likely out of sheer fear and reluctance) into the dark miasmal gumbo of hogmind. In fact, there are precious few such counselors who feel competent even to try—which itself probably goes a long way toward explaining the comparatively slow growth of hogology in this country.

Nonetheless, a certain Mr. Fred Kimball of Gardena, California, does manage to qualify as one of the nation's rare porcine psychotherapists, though even he acknowledges his own limitations and the general need to delve further. Mr. Kimball—who, at seventy-five, is active on the lecture circuit, in addition to teaching occasional classes in animal perception and appearing now and then on West Coast TV shows—undertakes most of his counseling in his office in Gardena, but is not at all averse to making house calls.

"I'm an old man in years," he beams, "but youthful in many ways." Kimball is an ex-Marine, a former merchant seaman, and for twenty-two years was a professional wrestler before he became engrossed in extrasensory perception and its application to the animal world, about which he claims to have written three books thus far. "I'm a psychic," he allows. "Most of my work is psychoanalytic. I work with people, too, because there's not much money in animals. Like every time I go on a show I entertain everybody, read all their animals, but I don't get to sell my books and I don't get paid for it."

Kimball is an enthusiastic, rapid-fire talker who speaks gently but with the flat edges of a lingering New England accent. "On shows, people bring in animals that I've never seen or talked to or met or anything else. I've had quite a few write-ups on this. They contacted me once about going to talk to a porpoise. The man asked me out and I canceled my schedule but he didn't show up. I've worked with hogs a little bit. Frankly, people just don't bring 'em. I've had horses, I've had guinea pigs, I've had rabbits, squirrels, I've had gophers, but no one's brought any hogs to any of the TV shows I've been on. I've gone out and talked with them on various farms but I haven't done any really deep readings on them. It's like a man brought a dog to me back in Cedar Rapids, and this dog had lived in a kennel all its life so it didn't have very much it could talk about."

Kimball says he establishes his initial contact with animals through the spirit world. "You look at the picture in their mind and try to get the picture that they have and then put words to it." Mr. Kimball is disarmingly humble about his semi-mystical ability and believes it's not all that rare a phenomenon. In fact, says he, "There's a Mexican out here who's able to send thoughts to animals. He'll tell a horse, 'If you come up to the fence right away, then tomorrow when I stop I'll have something good for you.' And the horse'll be there. He does it with all kinds of animals. But he says, 'I always lie to 'em and tell 'em I admire them and understand them,' and he says they respond to this."

Concerning his own office work, Kimball explains, "People come in with problems with their dogs, problems with their cats—about their life and what they think or what they feel. My real work is with people, but I work with animals because people want to know these things about their animals. Sometimes they just want me to read their animals' minds." His first contact with hogs, he says, was out of curiosity, "just to ask 'em a few odds-and-ends questions. I've asked about how their life was, who they fear, how they feel. Just like you'd ask a cat—'Well, who do you get along with, who do you hate?' The same thing with horses. You ask a horse, 'Who's your friend? How fast can you run?' And the horse will tell you why it won't run or doesn't like this or that. It's all about the same—just ordinary animal questions. I can't say that I've had any conversations with hogs that I could check out what they said with their owners. I've had a farmer tell me, 'Oh, no, you're not going to talk to *my* hogs.' Never even gave me a chance, see? But I just try to ask hogs about their life, how they feel about the others, their food and things like that. You could ask a prize hog, 'Did you do anything to bring this about?' And he'll probably tell you, 'No, I'm overfed or someone does this or that or they give me special treatment.' It's just ordinary chit-chat, don't you see what I mean?"

Kimball readily asserts that hogs are highly bright, "but again it depends on how restricted they've been and how much chance they have. I get out and prove by evidence. You take me to any place a man has his hogs and if he knows them as individuals then I could tell him, 'Well, *that* hog has this and this hog says that.'"

Now the prevailing psychological problem Kimball finds among canines is that they've usually had little time to be trained by their parents "so they start learning only from humans. Most dogs don't even know they're dogs. They want to eat candy and all these things humans eat that aren't good for dogs—or us humans either, as far as that's concerned." To a large degree this is true of hogs, too, he believes. "I remember when I was a kid back in New England we used to have hogs that were family pets. And I met one hog someplace that was a *watch-dog!* The man said it was better than any dog he'd ever had. It was a good one. It was quite smart. He said it would warn him and was more attentive and better than any dog because for one thing it wouldn't sleep as much, or as solid."

"Have you had many dealing with cows?" I ask.

"Oh yes," Kimball nods. "I went out here to Rockwell Dairy Farm. A fellow there, I taught him to box, Frank Stockton, he fought all the top fighters. Also, he usta dive a hundred twenty-four feet into a six-foot tank of water. He missed it and broke his back when the wind blew him sideways at the Frisco fair. I went out there to the dairy farm where he was handling this prize stock—Golden Guernsey stock. A cow there said, 'I had a baby and they took it away too early.' Frank looked at me and said that was right. I asked a bull about what prizes it had won and it told me it had won one first and two seconds. He'd been to San Diego and to New York. He didn't know the names of the cities; he said, 'I've been south and I've been east.' I usually go every year to interview the various cows and bulls. Oh yes, I've done a lot of work with cows."

As to the question of whether his counseling usually results in behavioral changes in his animal patients, Kimball replies that, yes, he's gotten lots of letters to that effect. "People have told me I improved their animal. But I think it's their own mind that changes and understands the animal better, and it's *that* that makes the animal change." Kimball's psychic conversational ability makes him something of a liaison between men and their beasts; and as he comes more and more to understand particular animals he is able to convey their personal hang-ups or attitudes to their owners. He understands cows quite well. "Cows, you know, don't mind dying when they're older. They think they have a right to be of service to humans. They have no compunctions about dying, no fear of the hereafter or anything else. They're pleased to be of service and all of this. But they don't think it's right for them to have to live in their own manure like humans make them do. These are the kinds of things you run into."

Hogology, however, still has a long way to go; for the hog himself still evokes uneasiness among many people who either fear him or perhaps close themselves to thoughts of those things grown down deep in the murky soil of the swine soul. Thus the hog tends to be viewed through a maze of hoary notions and legends. His generally odious (and undeserved) reputation remains lodged in the bowels of conventional belief. Nursery rhymes and childrens' stories first distort the porcine image and then continue to reinforce that distortion in young minds— while the clichés, metaphors, various linguistic indiscretions and bits of anti-hog invective on the part of adults do further disservice.

The necessary search for a fresh definition of hogness must begin, then, by summoning up some of society's dank old myths and pitting them one by one against newly acknowledged actualities. Most of these points you should be generally aware of by now.

To begin with, we are told—in fact, we are practically brought up to believe—that hogs are instinctively and rigorously *dirty*. Well, apart from what has already been asserted by on-the-spot authorities, even such a stiff-lipped source as the Encyclopaedia Britannica says, "Contrary to general opinion, the pig is a clean animal if given sanitary surroundings." Let us lay this notion to rest forever.

It is claimed next that hogs are *gluttonous,* if for no other reason that a hog is a hog and hogs are . . . well . . . gluttonous. Again, faugh! Gluttony is overindulgence and, as was pointed out earlier, hogs (unlike horses, cows and others) know when to quit. On top of that, as John

Beresford wrote: "But reflect, reader, how it would be with you if you had a barrel-shaped and capacious body carried on four very short legs: if you had a nose especially constructed and designed to go to the root of matters: if you had a mouth of peculiar capacity stretching almost from ear to ear . . . Would you not enjoy your food even more than you do now? Would you not grunt, and even slightly squeal, with the excruciating ecstasy of creamy, rich barley-meal, as it entered your mouth, gargled in your roomy throat and flowed into that vast stomach forever clamoring to be soothed?"[4]

Again, hogs are charged with *laziness and greed,* when actually they are industrious to the point of obsessiveness because, among other things, they are persistently inquisitive about their surroundings; and, too, they don't sleep even as much as dogs. Moreover, greed has no place in the mind of the hog, for his preferred style of life is essentially ascetic with an insistence on the priceless value of the simplest things.

And hogs, say members of the smart set, are *pigheaded.* The fact is, hogs, if anything, are simply not so weak-willed and pusillanimous that they'll blindly rush to do any idiot thing you ask of them. Though wisely circumspect, they constantly show themselves to be independent, innovative and adaptable.

Then too, hogs are reputedly *clumsy.* Well, no one who has ever watched a hog could say such a thing as this. Hogs are the essence of nimbleness and agility afoot. One never sees a hog trip, stumble or accidentally bump into things.

Lastly, a phrase familiar both in Greek and Roman literature maintains, "You cannot make a silk purse out of a sow's ear." Now up until recently this may have been true for all anyone knew. But on November 27, 1971, Mrs. Tinie Smith of Atlanta, Georgia—having experimented with different techniques in a manner as painstaking and lengthy as Edison's quest for the proper filament to use in a light bulb—perfected a genuine silk purse made entirely from the ears of sows, which was said by many of her neighbors to look "quite smart." Mrs. Smith says she hopes to refine the process even further and perhaps eventually manufacture what she calls the "sowlk purse" on a commercial basis. She says she's working on one now which she hopes to send to the President's wife as a promotional gimmick.

Altogether, as can be seen, the unchallenged conventions of daily life would sooner or later have buried the hog under tons of institutionalized silt were it not for such persistently emerging little triumphs and revelations as these.

If nothing else, the custodians of the old consciousness will eventually die out, but the hog, bless him, will persist. All the evidence is not yet in, but present indications suggest that the full story of the hog—what he can do and be—is going to bust a lot of balloons!

One major area of further exploration is the breeding of specialized swine. Over the course of thousands of years, a vast variety of dogs has been developed—through selective breeding, by luck, out of necessity or whatever—to serve man in a variety of highly specified ways. Burly dogs were bred as guards, bulldogs for combat, sausage-shaped dogs for burrowing after small game, pointers, sprinters, retrievers and so on. What progress has been made in hog breeding, however, has just been an extension of the single-focus *status quo*—just variations on the relative quality or quantity of the creature's meat. But the amazing thing is that even though this has been the sole intention behind breeding and cross-breeding the hog, he has maintained an extraordinarily keen mind and set of talents to match; he remains a creature who has highly subjective intentions and meaningful goals. His potential has been obscured by the narrow range of specialty for which he has been bred. To many hog-men this seems uncanny; it even appears to some as something deliberately contrived by the gods to vex and bedevil man and make him feel guilty. In any case, there is considerable cause to believe that over the years it has been precisely *because* of the hog's outstanding attributes that society has been afraid to let him evolve into whatever he might possibly become.

One potentially profound force which may alter the status of hogs (and hogmen) in relation to the rest of society is the burgeoning development of meatless meats. As things presently stand, Americans consume roughly 1.5 billion pounds of bacon per year. But now at last upon the horizon there has appeared a product that looks for all the world like bacon, comes in strips, doesn't shrink when fried, has no cholesterol and only about one third the calories of the real thing. More or less simultaneously there have emerged ham chunks, ham slices, sausage links and patties and pieces of pork—all of it made from soybeans, wheat and oats processed to look and taste astonishingly like hogmeat. These new concoctions, which are constantly being improved upon in the research departments of companies like General Mills, Miles Laboratories and General Foods, are made of textured vegetable protein—the end result of a new technique that gives soybean meal a meaty texture with, in most cases, even higher quantities of protein than is found in the meats they are intended to imitate. Soybean technology has reached the point where almost any meat can now be duplicated, and at less cost all around than under the old system of things. For instance: feed a steer on the meal extracted from 1,440 pounds of soybeans yielded by the average acre and you'll end up with roughly 58 pounds of protein. Process that same soybean meal into textured vegetable protein and the result is approximately 500 pounds of material that approximates the protein content of meat. And the price to the consumer is dirt cheap in relation to genuine animal flesh. As the president of Miles Laboratories said, "We want people to like these products enough to eat them in preference to other things." There are millions of creatures abounding across the land who would like that too.

If this does indeed come about—even on a partial basis—there's obviously going to be a monumental shake-up in the social role of hogs; there's going to be less of an urgent commercial need to breed hogs merely to eat. They can be free to fill all sorts of other functions. And there's no worry about there not being enough of them to go around, for hogs are so rigorously prolific that if a single sow were left to breed continuously, and if all her daughters did likewise, she could wind up in ten years' time with some 7 million descendants.

Many people feel this coming even now. There's going to be a massive redefinition of the hog for the new age—for this whole thing, after all, revolves around the question of his future identity. There is every reason to be hopeful. Indeed, the free and unfettered evolution of hogs may very well result in wonderful new creatures with which to cohabit the earth. In addition, there's the chance to promote controlled hog breeding from an entirely new point of view. "Again," wrote G. K. Chesterton, "we do not know what fascinating variations might happen . . . Types of pig may also be differentiated; delicate shades of pig may also be produced. A monstrous pig as big as a pony may perambulate the streets like a St. Bernard without attracting attention. An elegant and unnaturally attenuated pig may have all the appearance of a greyhound. There may be little, frisky, fighting pigs like Irish or Scotch terriers . . . Artificial breeding might reproduce the awful original pig, tusks and all, the terror of the forests—something bigger, more mysterious, and more bloody than the bloodhound . . . With elaborate training one might have a sheep-pig instead of a sheep-dog, a lap-pig instead of a lap-dog . . . Why do you not at once take the hog to your heart? Reason suggests his evident beauty. Evolution suggests his probable improvement."[5]

Now it must be pointed out here that these same projections of boundless potential could not in all honesty be applied to all other domestic animals, many of whom, while they may indeed have some room for growth, are simply more limited. Cows, for example. Certainly it is possible to conceive of a slightly more streamlined bovine body, but even if that were achieved the fact remains that your average cow still brings little or nothing to a conversation. Nevertheless, they, like other creatures, can show us certain preternatural phenomena. In the late 1960s a British scientist became curious about why it almost always happened that groups of individual cows idly grazing in fields all seemed to be pointed in the same direction. No matter how scattered they might be across a field or pasture or hillside, their heads always appeared to be aimed the same way. He pondered and probed away at this perplexing question. Perhaps they all instinctively headed into the wind? No, that wasn't it. Perhaps it had to do with the sun or the season or the time of day or the temperature? No, nothing correlated—until he eventually discovered that cows tended to point, almost like magnetic compass needles, in the direction of subsurface deposits of iron ore. This phenomenon, which he termed "Bovinity Flux," rates now as one more of those fascinating little signals we can receive from the animal world—in this case, the cow world—that enhance our understanding of life and our relationship with the earth itself. (More about the cow later on.)

Even today there are multitudes of people who persist in the socially instituted belief that there are precious few things a hog can possibly reveal to us, for which reason it is said he had best therefore be kept exactly as he is. Needless to say, new forms or proposed innovations in anything always appear immoral or chaotic; yet it is inevitable that at some point these hitherto untrod paths must and will be traveled. It's probably fair to assume, for example, that at first the idea of using pigs to hunt truffles seemed somehow bizarre and out of the question. But as it has come to pass, talented French and Italian pigs have been tracking down and rooting up truffles (which humans couldn't otherwise harvest) for centuries now.

Another lesser-known but proven pig talent is pointing and retrieving game. This first developed in England between the eleventh and the fifteenth centuries, at a time when only nobility were allowed to hunt. To help enforce this exclusivity, peasants weren't permitted to keep dogs over ten and a half inches high since those that were any larger could be used in illegal hunting. Thus, in place of dogs, the British peasantry discovered pigs could be trained to hunt and retrieve just as well.

In eighteenth-century England a gamekeeper named Toomer decided to train a certain sow to hunt. "The first step was to give her a name, and that of Slut (given in consequence of soiling herself in a bog) she acknowledged in the course of the day, and never afterwards forgot. Within a fortnight she would find and point partridges or rabbits . . .

and in a few weeks would retrieve birds that had run as well as the best pointer, nay, her nose was superior to the best pointer . . . She frequently stood a single partridge at forty yards' distance, her nose in an exact line, and would continue in that position until the game moved."[6] And finally, in Portland, Indiana, just a few years ago, farmer Jack Hough trained his pig Barney to point birds and rabbits. When Barney finally got too fat for field work he doubled as a saddle horse for Hough's son.

There are other instances in history of hogs replacing horses. Heliogabalus, the Roman emperor, trained boars to run his chariot; and in the nineteenth century, "a gentleman had trained swine to run in his carriage, and drove four-in-hand through London with these curious steeds."[7]

Altogether, as can be seen, when his potential isn't repressed, as it ordinarily is, the pig is a veritable Pandora's box of exciting possibilities. It is astounding to imagine all the things even ordinary everyday hogs could do—if they only just had the time. But the further realization of this potential requires us to declassify and redefine the hog today. And this task, in turn, requires the active concern of men and women who are not afraid to foray into these endless caverns, for hogs are constantly showing us mystical and subterranean visions that we don't usually want to see.

It seems safe to prognosticate that, if all the pieces fall properly into place, hogs henceforth will no longer be classified on the basis of meat-type and lard-type. Instead they may be categorized as either "utility hogs" or "pleasure hogs."

Utility hogs—those assigned to carry out certain practical day-to-day functions for man—would obviously be the first sort to be trained and developed since their journeyman activities would bear witness to the porcine potential and thereby pave the way for all manner of other breeds and permutations. Truffle pigs, hunting hogs and carriage swine have already been touched upon, but for further possibilities in the utility class, witness:

WATCHHOGS: As previously pointed out, hogs don't sleep as much as dogs and they can be, Lord knows, every bit as protective and ferocious, not to speak of loyal and sensitive to the natural balance of things. Consider the response of a burglar probing dark and furtively through a Manhattan apartment when he suddenly hears, rising perhaps from behind a sofa, an ominously deep, pharyngeal grunt followed by the unstoppable onrush of a three-hundred-pound boar!

SANITATION SWINE: The labor problem being what it is, and hogs being the way *they* are, it has been proposed that specially trained hogs be employed—or held in reserve in the event of possible garbage worker strikes—to reduce the ever growing accumulation of trash on city streets. The hogs could consume those leftovers and edible items that would otherwise breed germs. Possibly ways could also be perfected for them to clear away other rubble (like paper, tin cans, etc.) by pushing it into storm sewers or onto special scoops attached to trucks. Their talent for rooting and shoving things could clearly be applied here, in addition to their natural proclivity for fastidiousness.

SEEING-EYE HOGS: This is a very valid possibility. Hogs are, by nature, extremely considerate of others and certainly hard for drivers or pedestrians to overlook. They are not prone to drop everything and chase cats, and their heads' normal position casts their eyes in such a direction so as to spot rocks or curbs their masters might otherwise trip over.

SHEPHERD HOGS: Again, what with their non-aggressive tendencies toward other living creatures, coupled with their speed and air of authority when called upon, hogs would be ideal for keeping groups of sheep or goats in order—and perhaps teaching them a few things as well. There has already been some successful experimentation with hogs in this role.

MILITARY HOGS: Warfare quite naturally goes against the grain of the standard, peace-loving hog, but on the basis of their strong sense of duty it is possible to put them to use in a number of conceivable combat situations. Animal Behavior Enterprises, in fact, undertook a related study at the request of the U.S. government and discovered, for one thing, that hogs could carry a lot of weight. Also, since enemy troops would normally be on the lookout for dogs, hogs could be used to carry messages—"if," cautions Marian Breland, "they didn't get eaten *en route.*" A certain Colonel Wallace Tashmar (U.S.M.C., retired) has for years actively advocated the use of hogs in mine-detecting as well as the training of para-hogs, patrol hogs and tracking hogs. Colonel Tashmar claims that, in the event his suggestions are taken, he has already gone so far as to work out an entire manual of arms along with techniques of close-order drill for a typical hog infantry platoon.

It should be clear that the opportunity for training various types of utility hogs is limited only by the number of practical tasks that need to be accomplished. But it's important in all of this that people don't become locked into the notion of regarding their porcine friends only as creatures of labor, for they are legendarily entertaining and capable of great fun as well. It is in this regard, therefore, that the entire concept of "pleasure hogs" comes forth to represent the further fulfillment of swinedom. Along this line, those hogs we are already familiar with who are currently employed as carnival attractions, Thespians or acrobats are good examples of the "pleasure" type. But as for still more possibilities and variations consider:

RACEHOGS: This can be realized in a number of ways, partly on the basis of the fact that hogs are quite fast, and—either on their own or with a jockey—may gallop, prance or trot. This latter, of course, calls up the possibility of hog harness racing, complete with sulkies. The breeding of sleek racehogs would have the additional effect of creating a whole new physical image of pigs in the public's mind.

RODEO HOGS: Owing, once again, to their great flexibility of movement—not to speak of their innate disposition to entertain—hogs may be used for bucking, roping (either roping the hog or roping something else while astride the hog) or general stunt riding.

TENNESSEE WALKING HOGS: Dr. E. L. Ranzbottom, in his book *The Hog and Ewe*, describes the activities of some enterprising hogmen who coaxed certain of their hogs to cultivate that short, sharp, high-kicking stride once thought to be the exclusive talent of particular horses. Once in the ring, these hogs instinctively fell into a lovely smug strut with their tails held high and their heads cocked at an angle of great elegance.

One humane and very compassionate notion recently advanced and worth considering is that of placing properly trained and housebroken young pigs in old folks' homes where they "could lavish untold affection on each other that would otherwise go down the karmic drain." A spokesperson for the Gray Panthers, an organization of activist-minded elderlies, views this as an idea that would quite possibly raise society's consciousness in a positive way toward both old people and hogs simultaneously.

Once again, the attainable possibilities for developing hogs of pleasure are virtually infinite. It's just a matter of breaking through the public's resistance to unfamiliar concepts; yet the fact remains of the hog's emerging potential for greater freedom and fresh manifestations of soul. The barriers exist only in the minds of men. But improbable enterprises are often sustained by faith. And there is ample reason for faith in the creature, for it has long been realized that "the hog may be on to something." The initial breakthrough may possibly come with the significant symbolic decision to unring all snouts and let hogs evolve to take up, individually, whatever role suits them best. Each newly discovered function for the hog will, as it is realized and demonstrated, make all the previously acknowledged talents more clearly visible and obvious.

Perhaps a good place to begin would be with the wider acceptance of something that is already obvious to many hundreds of satisfied people:

the great pleasure to be found with pigs as pets. "I could never imagine why pigs should not be kept as pets," wrote Chesterton. "To begin with, pigs are very beautiful animals." Many others have concurred with Chesterton's perception, both before and since. Sir Walter Scott, for example, had a pet pig which he said had "never been severed from my company."

People who, for one reason or another, adopt pigs as domestic pets discover that they can be housebroken in practically no time, that they train easily, eat reasonably and develop genuine attachments for their owners. Then they go on to discover even more things about pigs. Writing in *The Atlantic Monthly*, a lady recounts the tale of her pet pig Henry, who ". . . sat down on command; gave me first the right and then the left hoof. He sat up . . . jumped through a hoop, and climbed up and down a small stile. When his daily lesson was over, he knew that he would be rewarded with a treat and a bottle of beer."

Another owner of a pet pig, this one by the name Bijou, recorded his observations for *True* magazine. Among them: "For a while she tried to chase the car when we drove off. But unlike the dog, she soon had sense enough to quit. Also unlike the dog, she is constantly doing something useful—digging for roots or exploring . . ." Then there's Jethro, a pet boar in Bradenton, Florida, who learned to operate a sliding screen door when he needed to get out. And Princess of Tarpon Springs, Florida, who goes for strolls on a leash. And Clairese, a lovely young Landrace pet of photographer Al Clayton who is perfectly orderly and mannerly in her habits and even washes in the family bathtub.

One could go on and on with similar stories—tales and anecdotes to add to your hog scrapbook—all of them having to do with people who, on the basis of simple contact, have entered into a whole new awareness of hogs. The writer Louis Bromfield, however, seems to have been torn between new and old hog consciousness when he declared: "Look at pigs over a fence but never bring one into your life, for when you put an end to his existence you'll forever after suffer from memories as cannibal and murderer."

It is worth observing here that though Bromfield accepts the notion of killing hogs for food he refers to this as the act of a "cannibal," of one who dines upon his own kind. The fascinating thing about this is that hogs have been used in science and in laboratory experiments throughout history precisely because of the uncanny number of common qualities they share with man. Leonardo da Vinci studied the cyclic motions of the pig's heart. The eighteenth-century British medical researcher John Hunter declared the pig to be the most useful of all animals for physiological studies. The great Russian physiologist Ivan Pavlov tried

True "hogritude"—the mystical essence and condition of being an actual hog—demands extended periods of meditation.

Here's how Mr. Hog eats his corn: He first peels the husk with snout and trotter, then rotates the ear, deftly nibbling its kernels. Wet corn often ferments in troughs, and hogs get drunk on it as often as they can.

An eligible boar and sow bill and coo in amorous foreplay, as (below) some adolescent shoats gaze on with envy and wonder.

Toward the shank of another lively afternoon, a Texas porker—all alone and bound next week for market—weightily contemplates the monumental just-so-ness of the hog condition. And grunts upon it.

experimenting with pigs but gave it up because of their reluctance to lie still and be quiet on a table; he concluded that pigs were "inherently hysterical." And South Africa's Dr. Christiaan Barnard, noted heart transplant pioneer, claims the hog may suddenly become vital to the human race—or, as he put it, "may become the salvation of mankind." He believes it will be more convenient to use pig hearts to transplant into humans because, as he has stated, "Strange as it may seem, in several anatomical aspects the pig is closer to the human being than any other animal. Its heart often is quite big. Even the ape's heart is not big enough to provide a human with circulation of blood."

And this gets into another important aspect of our exploration. The chief drawback to their use in science has been the fact that the common adult hog is simply too large for handling in most laboratories. But this has now changed. Starting about 1950, the University of Minnesota's Hormel Institute began a program of specialized breeding with the aim of developing a smaller animal for the purpose of scientific study. As a result, there now exists the "mini-pig," which weighs about 140 pounds fully grown, or approximately the same weight as a man. The reason for going to all this trouble is that, as *Scientific American* magazine put it, "In anatomy and physiology the pig is remarkably like man." Outwardly, the body size and skeletal mass of a mini-pig are about the same size as a man's. Its teeth, too, are approximately the same size, which makes it a good model for dental studies. Also like man, the pig has comparatively little hair on its body and its skin closely approximates the properties of human skin, including the ability to give readings on allergic reactions, similar susceptibility to skin disease and even occasional development of dandruff.

Internally, the hog's heart and coronary arteries have much the same pattern as man's, its blood-clotting mechanisms are the same and it can similarly get a form of atherosclerosis. Hog and human bodies require much the same kind of food, they digest it the same way, and both have the propensity for developing peptic ulcers as a result of tension and confinement. Says *Scientific American:* "The pig's alimentary tract and metabolism are so similar to man's that the pig provides a standard for the feeding of infants and young children. It has been found that a young pig has more stringent food requirements than a human baby; consequently one can be sure that a diet that provides healthy growth in a piglet will be adequate for a baby." An additional discovery is that secretions from certain of the hog's glands are able to stimulate humans.

In terms of habits, both hog and human are fairly unique among animals in their willingness to eat anything, as well as in their tendency to grow fat and sedentary. Hogs respond to drugs like men and also have a personal fondness for alcohol. In truth, whenever and wherever they can, hogs drink. Sometimes the combination of heavy rain and heat upon shelled corn in feeder bins will cause the corn to ferment, much to the delight of hogs, most of whom are awesomely partial to demon spirits and, once inebriated, react pretty much the same as drunken humans. Some squeal gleefully, others get ill-tempered and still others stagger off somewhere and go to sleep. More recently, researchers at the University of Missouri-Columbia have discovered yet another point of resemblance —that the drinking habits of pigs, like those of many humans, seem to be related to their social status. The scientists monitored the drinking propensities of a pen of seven miniature pigs as part of an ongoing study of alcoholism. The team first determined the pecking order of the pen by noting the order in which the pigs lay down: the dominant, or top pig, inevitably lies down first in the corner of the pen, followed by the others in descending order. Then the scientists allowed the swine to drink their fill of screwdrivers (alcohol and orange juice), which previous research had shown to be a great favorite at porcine cocktail time.

The effect on the pecking order was spectacular. "The top pig drank so heavily that he lost his status within twenty-four hours," reported bio-chemist Myron Tumbleson. "The No. 3 pig drank very little and became top pig." However, the No. 1 pig redeemed himself the morning after and was back on top within seventy-two hours. Said Tumbleson, "He never became inebriated again." After that experience, the pigs settled into drinking patterns that were apparently determined by their feelings about their social status. "The heaviest drinker was the pig that ranked No. 6 in the seven-pig social order," explained Tumbleson. "Apparently he is frustrated about his position and has resorted to drink." But the low pig in the pen apparently felt no need to drown his sorrows. "No. 7 knows he's last," said the researcher, "and has accepted that."

As for putting some of these hog and human similarities into practical application, doctors are currently using swine for such things as research into the radiation treatment of cancer, the study of possible cures for rheumatism and arthritis, and tests involved with understanding the aging process. In the spring of 1971, the life of a thirty-one-year-old London housewife was saved by a pig's liver, through a process called "extra corporeal porcine perfusion." The woman's blood supply was diverted from her diseased liver and passed through the pig's liver for five hours, during which time her own liver was able to start repairing itself. In still other operations, the valves of hog hearts have been transplanted into humans. This is medically termed a "hetero-graft," or transplanting the tissue of another species, and it has been completely successful thus far in Russia, England and New Zealand. Dr. Edward T. Lee, who performed the New Zealand operation on a young woman critically ill with heart disease, said, "I wouldn't want to injure Adeline's career. You know how foolish some people are about this sort of thing. They might not want to employ anyone with a pig's heart valve in them. Of course, it's ridiculous."

The interchangeability of heart parts clearly suggests an astounding measure of physical closeness, if not spiritual intimacy, between hogs and humans. And maybe even more than that. Maybe, on a deeply subconscious dimension, we identify in some measure with them. Maybe, as Dr. Dennis Sikes, a research professor at the University of Georgia, says, "Man is more nearly like the pig than the pig wants to admit." Maybe— leaving aside all the social obstructions—man is essentially a hog beneath the skin. And this, in turn, leads to something else.

Biological explorations are continually revealing more and more about the fascinating substance called RNA, the genetic chemical that transcribes the inherited potential contained in DNA (the genetic memory molecule) into the proteins that compose the protoplasm of animal cells. In other words, RNA encodes and carries information, making it, therefore, the key to memory formation. Experiments with RNA have, in fact, demonstrated actual chemical memory transfer. Michigan psychologist James McConnell has shown this with Planaria, or flatworms. He conditions them by electrical shock to contract when a light is flashed. Then he grinds the worms up and feeds them to untrained worms, who are then able to learn to contract twice as fast as their predecessors. What happens is that the first group of worms form new RNA which molds new proteins containing the message that light is a signal to contract. Then the second group, having consumed the memory proteins, don't need to manufacture so much of their own; they have swallowed memory, so to speak. The same kind of experiments have been performed with rats wherein they are taught to fear darkness, after which their brains are ground up and injected into mice, which then react to darkness the same way. RNA, therefore, chemically converts experiences into learning which can then be transferred to the cells of another creature. All of which has led McConnell to speculate whimsically: "Why should we waste all the knowledge a distinguished professor has accumulated simply because he's reached retirement age?" Instead, McConnell proposes, the students should eat the professor.

It is clear from all of this that for thousands of years now humankind has been busily consuming not merely hogmeat but the *characteristics of basic hogritude.* This may partially explain the almost mystical attraction people have toward hogs. But it also offers an obvious means of boosting the porcine potential in a manner so simple and swift as to make up for our years of neglect. The brains of people recently deceased (and chock full of all that human learning in its chemical form) should immediately be fed to hogs. Even as things now stand, the hog appears in the eyes of many as some amorphous, vaguely kindred beast containing unknown hosts of energies and impulses. But once he has consumed human RNA, he could begin to attain his real fruition and, too, become clearer to us in the process.

There have been scores of reports thus far—though none documented —of attempts by husbandmen to interbreed with their sows, presumably in the spirit of a scientific effort to enlarge the swine horizons. The problem here, of course, has been that hogs and humans don't possess the same number of chromosomes. But perhaps, in time, ways will be found to overcome this biological obstacle. In any case, these *ad hoc* interbreeding efforts do seem to be continuing.

Meanwhile, and on a more modest scale, it may well turn out that the newly developed mini-pigs—who won't outgrow the cost of their upkeep—will be the inspiration for redefining the whole hog-human relationship. It may well just be a question of size. A big thing, after all, becomes frightening; particularly if it's a *smart* big thing.

But in whatever way the breakthrough is made, the ultimate aim remains the same: to permit the boundaries of the hog-self to expand and encompass wider identities taken from the surrounding culture. No longer could society then continue to remain disdainful, citing as an excuse all the ingredients that don't go into hogs. For slowly, even now, the hog is coming into his own.

Yet for the time being, the hog, alas, remains in bondage to the obduracy of officialdom. His life remains built upon a narrow base, while each small glimmering of potential reinforces, in some people, the apprehension that hogs are somehow a threat to man—a fear thereby enlarging upon society's reactionary and perpetual piglash. Putting down the pig, however, is chiefly a habit of unthinking; and it is one which tends to work to the detriment of us all. For out in the deepest boondocks right now lie millions of piglets waiting for their future, each already pregnant with portent and good things to offer. Yet in the eyes of the prevailing system, piglets are merely part of a series of inexorable continuations. If you've seen them all, you've seen one. They are victims of technique, destined for a canned-food world and cut off from all personal sensations or long-range hopes.

But discovery makes old forms of perception visible and places them in perspective. New discoveries create new and larger audiences, and a fresh sense of perspective permits people to recognize the fallacy of old patterns. All of this offers hope for declassifying the hog—for removing him from his traditional niche and placing him in a fresh context in people's minds. With a change in the hog's role, a change in men's attitudes will surely follow. Perhaps it may simply mean a return to the outlook held by people of earlier times, such as the British peasantry, among whom "the hog was, in a manner of speaking, one of the family, often seen following his master from place to place, and grunting his recognition of his protectors."[8] Or perhaps it may mean the acceptance of a more valid place for multi-talented individual hogs among us. Or maybe the recognition of an entire swine subculture in our midst.

The hog, merely by being who he is, provides an enormous potentially creative force for changing his own environment by changing the minds of the people who shape it. Obviously this stands to improve the weal of all hogkind. But more than that, it may presage a quantum leap in our dealings with all other species of creatures.

Other animals that people commonly raise as part of their households —dogs, cats, songbirds, monkeys, turtles, rabbits *et al.*—may indeed be regarded with great fondness, but are still seen primarily as decorations and are still relentlessly humiliated. Many pet owners tend to think it quite smart to bedeck and festoon their captive critters with ribbons and bibbons and fancy collars and little jackets, or to fix their hair in permanent waves, clip their ears, bob their tails or paint their toes—all of which are badges of debasement and suppression. Such people regard everything as a metaphor for man. Human emotions are attributed to the acts of even the simplest animals. If a parakeet comes forth with a melodious song at the close of a bright spring afternoon, he must be reveling in the joy of being alive on such a day. If a female cat carefully washes her kittens, it's because she wants to show her babies the depths of her motherly love. If two male dogs do battle in the presence of a female in heat, it's because each is wildly jealous and also wants to display to her his bravery. Bah! The animal world doesn't run on human sentiments. Anthropomorphism is the classic error in terms of trying honestly to understand other species, and is the kiss of death. In view, therefore, of the way men presently treat their kindred creatures, the question is: Do we ultimately result in humanizing animals or in animalizing ourselves?

In mid-September of 1916, the Sparks Brothers Circus hit east Tennessee, complete with its dazzling entourage that included Old Mary, the star elephant. Old Mary had been feeling a little irritable and out of sorts earlier on the day of her arrival, and at one point during a parade down the main street of Kingsport, Tennessee, she suddenly stopped, hoisted her trainer with her trunk and dashed him to his death on the pavement.

The killing immediately became a civic and commercial crisis. The circus people, with their concern for ticket sales, and the town leaders, out of respect for public opinion, huddled together and came to the judicious decision that the elderly pachyderm should properly pay for the crime with her life. The problem, however, was that there wasn't a gun anywhere around of large enough caliber to do the job. But since the circus was scheduled to appear the next day in Erwin, Tennessee, someone came up with the suggestion, which was immediately accepted, that Old Mary could be hanged there with the wrecking derrick of the Clinchfield Railroad.

Word of this upcoming phenomenon spread throughout the area to the extent that over five thousand people from all around east Tennessee turned out the next day in the Clinchfield yards to witness the official elephant lynching. Late in the autumn afternoon Old Mary, led by one of the circus hands whom she considered a friend, lumbered into the yard and to the place of her execution. She stood peacefully as the circus people circled her neck with a huge chain and hooked the other end to the derrick.

The cables tightened, tightened and the wide-eyed crowd gasped as Old Mary was slowly lifted into the air, wrenching and twisting at the end of the steel noose. She quickly realized this wasn't just another circus trick and commenced now to thrash and struggle so frantically that the derrick swayed. Then a link in the chain snapped and the elephant fell to the ground with a huge crash, scattering the spectators in all directions.

Being understandably befuddled and outraged, she required considerable gentle coaxing to be lured back beside the derrick for another attempt at carrying out her sentence. Again her neck was placed in the heavy metal noose, and again she was hoisted into the air. This time, though, she didn't fight the chain; she came forth only with a brief, meek trumpeting and a single final flailing of her great gray leathery trunk and, moments later, quietly died from hanging.

Old Mary, prize pachyderm and stellar performer of the Sparks Brothers Circus, was laid to rest later that cool fall day in a large grave on the Nolichucky riverbank just a little ways west of the Clinchfield Railroad yards.

VIII
Hoglore

> . . . I am ashamed to tell you,
> But I will tell—I had bristles sprouting on me,
> I could not speak, but only grunting sounds
> Came out instead of words, and my face bent over
> To see the ground. I felt my mouth grow harder,
> I had a snout instead of a nose, my neck
> Swelled with great muscles, and the hand which lifted
> The cup to my lips made footprints on the ground.
> —Ovid's *Metamorphoses*

Far from the humble sty or hoglot, far removed from commercial pork-producing considerations, far from the mental reach of those who regard him with distaste or ignorant indifference . . . above, beyond and completely abstracted from all these matters, there glimmers a dominion wherein swine take on a wholly different and independent reality: the swirling realm of hoglore, the intricately interwoven land of legend, ritual, myth and . . . the Pig Symbolic.

Throughout history and down the throat of time, the hog has been one of the most persistently transcendent symbols in human culture, oftentimes raised to canonical and monumental form as a bearer of multitudinous tidings and as an all-around expander of spirit signifying, at once, the sweetness of life and tenderness of death. In spite of the hatreds and passions even his name inspires in certain circles, he subjects himself to us in good faith unendingly. For he remains a creature of fundamental gentility and courtliness and honor, and, as such, he is more often than not viewed as a totem of rectitude and a reminder that we are all moved by the same fierce forces. In one way or another, throughout the ages, throughout the cultural mythologies that have washed across the face of the earth, hogs are cast as agents of *liberation*.

On this spiritual and elevated plane—even more so than in the course of what presently stands as ordinary life—hog and human ceaselessly converge, merge, interweave and reach rapport.

At one time or another, hogs have meant all things to all men—or, at the very least, a lot of things to most men. They are miracle, fantasy, jest, nonsense and fable; they are objects of widespread reverence and deeply pious thoughts; they have penetrated like a wormhole into all moral phenomena and constitute a scary force in the course of human life; and they have been rated as semi-celestial creatures who provide a constant source of insights as to what's going on, or else what's *about* to go on.

Many people from all sorts of diverse cultures have used a hog to form or change an idea. For example, certain pig parts have been said to be endowed with mystic curative powers, as well as with the ability to impart certain character traits to those who consume them under the proper ceremonial circumstances. Hog auguries were mulled over in the most toplofty solemnity by early Romans, who placed faith in "haruspication," the foretelling of coming events by studying the arrangement and size of a sacrificial pigs entrails. The Roman Senate, in fact, commenced each legislative session with priests sacrificing a pig and stirring among its internal organs for omens. Certain people in Southeast Asia practice this even today. On another level of augury, "Pigs see the wind," as Samuel Butler once wrote; and it remains a fact that hog actions are still read by some as weather prophecies.

Priests and visionaries, of course, almost invariably hold the hog in high esteem. But in the cryptic recesses of the general tribal mind the hog is lodged far deeper still as some great and galumphing life-force which men have alternately chanted to, incanted over, raised up taboos against and stood in overwhelming awe of. They are every bit as mystical as unicorns—and infinitely more accessible.

In cultural mythology the swine which first spring to mind are those legendarily fearsome boars who were pitted against various classical heroes, such as Heracles who vanquished the Erymanthian Boar, Meleager and Atalanta who took on the indestructible Calydonian Boar, and Theseus who slew the Wild Sow of Crommyon. There's more to this than immediately meets the eye. These are more than simple Greek and Roman tales of godly exploits which happen to involve hogs, for they each seem to spring from a common origin in ancient Greek religion. The name for the Greeks, or Graikoi, came from the fact that in pre-Homeric times the people originally worshiped the gray Crone-Goddess, who was most often represented as a sow. The latter-day stories of these hog killings, therefore, can be seen as symbolic of the suppression of this

earlier religion when the sow-goddess cult was supplanted by the more benign gods of Olympus.[1]

The rituals held for this crone sow-goddess "were in spirit dark and full of dread; the offerings were not of cattle, gracefully garlanded, but of pigs and human beings . . ." The annual spring rites included an "un-Olympian type of sacrifice—a holocaust of pigs—carried out, as one Greek commentator observed, 'with a certain element of chilly gloom.'"[2]

According to Cretan legend, Zeus himself, in one of his earliest manifestations, was said to have been suckled by a sow (probably the sow-goddess) and was later killed by a boar. As a matter of fact, being killed by a boar appears to have been an occupational hazard among ancient gods. Adonis, who was first worshiped by the Syrians and later by the Greeks, was killed by a Wild Boar while hunting, as also was Attis, a Greek god of vegetation. Under similar circumstances, a boar was supposed to have killed the Egyptian god Osiris, who personified the corn-spirit. In all three cases, the devotees of these gods religiously abstained from hogmeat. And in all three cases, legend hints at further and still more intricate mystomagical connections between hogs and these particular gods, in addition to numerous other deities.

One such fairly familiar god-hog linkage has to do with the Greek goddess Demeter and the rites surrounding her and her daughter Persephone. As the story goes, Persephone was out playing in a field one day when she spotted a specially fine patch of flowers (which actually had been sent up by Hades, Lord of the Underworld, to seduce her). When she ran to pick the flowers the earth gaped open and carried her down to become Hades' queen. When Demeter set out looking for her she found that Persephone's footprints had been obliterated by the tracks of a pig.

The ancient ritual commemoration of all this was a three-day festival called Thesmophoria, during which live pigs were thrown into sacred caverns and left to rot for a year, while the leftovers of last year's pigs were brought up and placed on an altar. Whoever got a piece of decayed flesh and sowed it with his seeds was guaranteed a good corn crop.

The deeper connection, however, is this: Demeter was the corn deity to whom the pig was sacred; in art she was portrayed with a pig in tow or carrying a pig or something like that. Originally, though, according to Sir James G. Frazer in *The Golden Bough,* "The pig was the embodiment of the corn goddess herself"—and those pig tracks found at the scene of her disappearance "were the footprints of Persephone and of Demeter." Moreover, since no image of Persephone was thrown into the caverns during the Thesmophoria, "we may infer that the descent of pigs was not so much an accompaniment of her descent as the descent itself, in short, that the pigs were Persephone."[3]

This same kind of interchangeability between hog and god originally applied as well to the other deities mentioned—the ones who were supposed to have beeen killed by boars; for "it may almost be laid down as a general rule," says Frazer, "that an animal which is said to have injured a god was originally the god himself." Attis, Adonis and Osiris were all three worshiped as deities of vegetation and fertility, all three were celebrated as gods of death-and-resurrection who regularly rose from the dead each year, and the worshipers of all three broke their general hog-meat abstinence only during their annual rituals when "the pig was slain as a representative of the god and consumed sacramentally. Indeed, the sacramental killing and eating of an animal implies that the animal is sacred, and that, as a general rule, it is spared."[4]

The idea of pig divinity is supported all the more strongly here by virtue of the fact that the pig was eaten annually by the worshipers of Osiris in Egypt—a land where, ordinarily, the pig was regarded as unclean and untouchable. If an Egyptian even so much as brushed shoulders with a passing pig he was supposed to step into the river fully clothed to wash off the taint. This, though, says Frazer, "also favors the view of the sanctity of the animal. For it is a common belief that the effect of contact with a sacred object must be removed, by washing or otherwise, before a man is free to mingle with his fellows" who apparently "conceive of holiness as a sort of dangerous virus."

Insofar, then, as everyday pig-touching is concerned, the Egyptians' beliefs imply that the animal was looked on not as a filthy and disgusting creature but as a being endowed with high supernatural powers, and that as such it was regarded with that primitive sentiment of religious awe and fear in which the feelings of reverence and abhorrence are almost equally blended. What is divine is simultaneously hazardous.

In this kind of uneasy equilibrium, one or the other of these contradictory feelings is likely to prevail. In this case, Osiris eventually became anthropomorphic and his original relation to the pig was forgotten. And once this came to be so, people also lost track of the fact that the pig was unclean on account of his holiness, figuring instead that if he was unclean it must be because he's just unclean. Hence the official rationale for sacrificing a pig to Osiris was that the creature was the god's enemy.

Still, a consistently recurring figure in the earliest myths of ancient folk is that of an Earth-Mother goddess who is associated with the soil, planting, fertility and with human sacrifice, and who—more often than not—is represented in porcine form. The Roman historian Tacitus wrote of the Aestii, a Germanic tribe, who worshiped the mother of gods whose emblem was a boar, for which reason boars were sacred to them. Then too, the people of ancient Gaul worshiped a goddess, whom the Romans equated with Diana, who is pictured "riding a Wild Boar, her symbol and, like herself, a creature of the forest, but at an earlier time itself a divinity of whom the goddess became the anthropomorphic form."[5]

A first-century Celtic deity, with jurisdiction extending into parts of Britain, was a boar-god depicted with the "eyes of the Great Mother engraved along either side, who, as lord of the wilderness, the underworld and the vital forces of nature, was also king of the land below the waves."[6]

Irish mythology maintains that one of the first arrivals in Ireland was Banba, a goddess who was sometimes said to be one of the daughters of Cain. Ireland was named for Banba—whose name means "pig."[7]

And in India, the Hindu mother-goddess Kali, who gives birth and death to all beings in the universe, has been represented as a black sow that sends forth her creatures only to consume them. A Buddhist sutra from the Ming Dynasty, about 1520, illustrates the Hindu trilogy of Brahma, Vishnu and Siva borne through the clouds in a chariot drawn by hogs.

Somehow there's a whole constellation of correspondences and features common among ancient pig-related mythologies even though they were independently conjured up and practiced in totally unrelated parts of the world. All of which suggests either a certain universality in the most deeply moving messages men have read from hogs, or else the existence of an original mystic communal bond between hogritude and humanity. For example, in legends, rites, sacrifices and various other forms of spiritual carryings-on around the world, hogs and humans have been viewed with perfect interchangeability. Moreso than any other animal, hogs have been regarded and employed as a substitute for—or at least as a creature most closely identified with—man.

In the mythology surrounding Demeter and Persephone—as well as in various other Greek rites—the use of pigs as sacrificial stand-ins for humans shows precise analogies with rituals of Melanesia and the Pacific. The general way these things are regarded is that a magical power is gained according to the measure of one's sacrifice. The best possible sacrifice, of course, is one's self. Next to that is another human being (a son, slave, prisoner of war or whatever). After that, the most potent offering is a beast of a species mythologically associated with the idea of divinity. And in most cultures this has usually meant the pig.

On the island of Malekula in the New Hebrides a series of pig-sacrificing ceremonies called the Maki is celebrated among megalithic shrines that resemble those of prehistoric Ireland. When a Malekulan dies he must reach the land of the dead by way of an entrance guarded by a goddess to whom he is obliged to offer a pig for her to eat in place of himself. This is a boar he has specially bred for the purpose, with tusks that form a circle representing the waxing and waning of the moon. "The boars, then, are the moon at the moment of its death consumed by the guardian of the underworld. Their tusks point to the continuance of life. Without these tuskers the man can neither enter the land of the dead nor be reborn."[8]

Other cases of hog and human reciprocity involve the West African tribal wizards of Gaboon who mingle blood with their animals to establish such a mythically intimate union that the death of one entails the death of the other. Also, claims Frazer, the Balong of the Cameroons believe they each have two souls, one of which may dwell in an animal such as an elephant or a wild pig whose death would cause the death of whatever human whose soul it possessed. On the island of Wetar in New Guinea many people believe that they are descended from wild pigs and that if they eat pigflesh they'll go mad. Then too, tribesmen inhabiting the Andaman archipelago south of Burma believe a mythical goddess of theirs turned some of their ancestors into pigs, for which reason pigmeat can be eaten only under ceremonial protection.

The aspects common among these rites and beliefs are in the associations of the pig with death, the underworld journey and with the mysteries of immortality. In the Melanesian ceremony of the Maki the pig is a human's direct counterpart who serves as an opener of the way and guide to everlasting life beyond death—just as in the ancient rituals of Greece, Egypt, Rome and other civilizations the pig was the sacrificial creature whose recurring symbolic meaning revolved around the portentous theme of Death and Rebirth. One way or another, pig means imperishable.

This theme occurs not just in ancient rites and ritualistic activities but also in literature and still-functioning folklore. In the *Odyssey*, for example, Circe drugs Odysseus's men and:

> *Instant her circling wand the goddess waves,*
> *To hogs transforms them, and the sty receives.*
> *No more was seen the human form divine;*
> *Head, face and members, bristle into swine.*

Then, after being prevailed upon by Odysseus, she alters the men once more, until:

> *Now, touch'd by counter-charms, they change again,*
> *And stand majestic, and recall'd to men.*
> *Those hairs of late that bristled every part,*
> *Fall off, miraculous effect of art!*
> *Till all the form in full proportion rise,*
> *More young, more large, more graceful to my eyes.*

This encounter between a hero who had 360 boars at home and a goddess who turns men to swine and back again, fairer and taller than before, underscores once more the connection between hogs and the idea of being reborn into some shape far superior to the original version. According to a similar bit of lore still active among the Irish, the ancient hero Oisin—one of the sons of the great giant Finn MacCool—was pestered for some time at his palace by a supernatural female with the head of a pig. Eventually she made herself known to him as the daughter of the king of the Land of Youth. Her father, she said, had cast her under a Druidic spell and given her a pig's head that would vanish only if one of the sons of Finn MacCool would marry her. In a true spirit of gallantry Oisin accepted the challenge, whereupon the pig's head was promptly replaced by the girl's natural beauty and the two then went on to become the happily-ever-after king and queen of the Land of Youth.

Still another variation on the theme of rude hoghides covering inner grace and radiance is the British legend of Sir Parzival and the Loathly Damsel, a wandering female who featured piggish bristles and a boar's snout. Those who lived lives steeped in sin saw her as a creature of unbelievable hideosity, while to men of "unsullied virtue" she came across as the essence of lovelitude. The basic notion behind stories of this sort is that any dolt instantly knows that the appearance of, say, an angel on the doorstep can mean only good things; but a similar visitation by a *hog* (or something appropriately hoglike) is something whose significance only the wise and holy-minded can recognize right off. In this sense, the hog stands as a summoner for a whole inward world of potentials unrealized in the visible order of one's time.

The point that keeps cropping up in all these mythical parables is that there's far more to the hog than meets the eye: that virtue, pulchritude and things of transcendent wonder abound beneath the porcine surface for those courageous or sagacious enough to risk a peek. Another aspect that comes across in the ancient hoglore of most cultures is the fact that the various mythical pigs—even including those who rate as gods or substitutes for gods—are not necessarily thought to be powers in themselves, rather they are signs through which the powers of life and its revelations are recognized and released: powers of the soul as well as of the outer world. Everything rises, flows and convolves together, and the hog ranks as an integral part of it all. He is a mythical mover and shaker . . . a catalyst . . . a constant metaphor for everlasting fertility . . . a down-to-earth apotheosis of higher reality: altogether, a creature imbued with an invisible purpose of uncommon force.

Symbols and myths are means of discovery that can reach a far greater reality than facts. They set up a certain structure in our relation both to nature and to our own existence. By expressing all manner of inner truths, the Hog Symbolic enables humans to experience even greater reality in the actual world. He is one of those abstruse elements, all full of contradictions and infinite subtleties, that drive the essential machineries of our existence. In fact, in any of its mythic, symbolic, ritual—or, so far as that's concerned, even fleshly—manifestations, hogness gives mankind a sense of itself in perspective, and thereby helps animate all human life. For, among other things, the lone hog himself is also an endless challenge: However he may be viewed, he still and everlastingly exudes a secret twinkling of all the bottomless mysteries of the irrational.

This is why hogs, almost simultaneously with their adoption as articles of livestock, became the focus of philosophical and religious ruminations, and why they have remained so. Few other animals can make that claim. Even Christianity is rife with hoglore. Representations of the creature pop up in churches high and low: sculpted hogs, soaring hogs, stained-glass hogs, hogs in relief, on bench ends, choir stalls, walls or roof beams. And they all mean something. "The pig Ecclesiastical sometimes finds its way into cathedrals through association with certain saints, most frequently with St. Anthony, the patron saint of swineherds."[9] But pigs also appear, particularly in older British churches, as illustrations of moral points or as satirical statements of certain practices or as caricatures of some ecclesiastics of the time. Or, of course, they may enter the church as heraldic emblems of various families whose coats of arms feature what are most often highly fierce and righteous (and hence very Christian) boars.

Usually the hog's Christian connections have had to do with symbolizing prosperity, creativity, humility and pureness of heart. But in the Middle Ages (which were lean times for hogs in general) they came to be depicted in terms of earthly vice and were even employed as agents of religious ridicule or symbolic malice. Henry VIII had his writing paper watermarked with a hog wearing a tiara to show contempt of the Pope. And churches in eastern Germany similarly used pigs to attack Judaism by way of carvings showing Jews sucking milk from the sows they officially scorned. During this period—surely hogdom's darkest era— swine signified sin, foolishness, fleshly lust, foul thoughts, greed, heresy, cruelty, backsliding and dirt. Also the devil. In Norman sculpture the hog stands for the forces of evil opposing Christianity—the "wild boar" of the 80th Psalm that roots up the tree of life. (Residual overtones of this medieval hog-scorn show up even in modern literature, as, for instance, *Lord of the Flies*, wherein the title role is played by a decaying hog's head impaled upon a stick and implied as satanic symbolism for social corruptibility and the ultimate plummet into evil.)

Probably the most blatant biblical association of hogs and evil spirits is the New Testament tale of the young man from Gadara whom Jesus came upon one day living wild and naked among the tombs—being possessed as he was by a whole bevy of devils. According to Mark, ". . . there was there nigh unto the mountains a great herd of swine feeding. And all the devils besought Jesus, saying, Send us into the swine, that we may enter into them." So Jesus forthwith cast the devils out of the man and into the two thousand swine which "ran violently down a steep place . . . and were choked in the sea." (As a matter of fact, in the forefeet of hogs even today, as you can see when the hair's been removed, there's a very small hole with six rings around it which looks altogether as if it had been burnt into the skin. The traditional view is that the devils entered by way of these holes and left their clawmarks. Anyway, there's no other explanation for it. And as if there were not already enough reasons for abstaining from hogmeat, Dr. Marvin H. Cohn, a pig-wise physician in Atlanta, declares that partaking of this part of the pig in particular "would be a *real* brush with eternity.")

It appears, though, that the real point of the Gadarene swine story is not that hogs and devils are naturally connected but that hogs and humans have a sort of spiritual substitutability for one another. And, too, this account is yet another legendary instance of human rebirth and transformation, with hogs as the essential catalytic agent.

Once again this underlines the hog's inborn sense of sanctitude and his transcendental cast of soul. In the unscaled eyes of men throughout the ages the hog betokens ultimate joy and spirituality, snout to tail. Observations on this are endless, and keep on popping up. As Huck Finn noted: "There warn't anybody at the church, except maybe a hog or two, for there warn't any lock on the door and hogs likes a puncheon floor in summertime because it's cool. If you notice, most folks don't go to church only when they've got to; but a hog is different."

The truth is, even dead a hog bodes more religious significance in *one single trotter* than, for instance, a live horse holds in his entire body. Surely there is no other living creature that so many groups of people of such widely divergent beliefs refrain from eating *solely for reasons ·of spirituality*. Among others, the Syrians, Egyptians, certain South American Indians as well as African natives, all Moslems (including, with especial vehemence, Black Muslims) and, of course, Jews have strict taboos about pork. On the basis of biblical commandments, Jews shun not only pig-eating but also pig-touching. This notion comes from the books of Leviticus and Deuteronomy, which declare as forever socially unacceptable and off limits any animals that don't manage both to chew their cud and have cloven hoofs—all of which is just a fancy legalistic way of saying hogs. The rationale is that the hog, by virtue of the fact that he is a hog, is "unclean." But to the Jews—as with the most ancient Egyptians—the hog's uncleanliness was originally a by-product of being extremely sacred. As Frazer states: ". . . originally, at least, the pig was revered rather than abhorred by the Israelites . . . Down to the time of Isaiah some of the Jews used to meet secretly in gardens to eat the flesh of swine and mice as a religious rite . . . as the body and blood of gods. And in general it may perhaps be said that all so-called unclean animals were originally sacred; the reason for not eating them was that they were divine."[10]

Everything fundamentally rolls and flows together. Varied beliefs which seem on the surface to clash are all, at bottom, really rooted in the same stuff, the same human perceptions and presumptions. So where, then, do the tides bear us? Here: to the core of the bond between hogs and the human spirit. On psychic, mythic, symbolic, theological or astral dimensions (or any combination of these) the hog, sacred or profane, persistently emerges as something that smacks mightily of mankind's spiritual alter ego. For the most part, men maintain a subliminal love-hate relationship with hogs, in much the same tenuous balance as that queasy juxtaposition between the sacred and the unclean. There remain, however, great gaps in the conscious acceptance of this natural bond and many visions left unglimpsed. The hog's flesh may be eaten, but he himself is not fully digested: he remains a repository for our fears of ourselves, an enigma formed out of our deepest uncertainties, the brunt of our self-reproach.

Perhaps, though, this all relates to the role of sainthood to which some cultures have relegated him. Men have always had deep intimations of the "inner hog." It could well be that hogs actually perform some priestly function by taking upon themselves all shades of human vileness. Perhaps they are ordained by the cosmos to do this for us silently. And then too, in silence, die for our sins.

Such are the gossamer speculations myths are made of; and until they are resolved in some other way they will continue to endure. Modern hoglore, in fact, includes overtones of ancient notions colored by all manner of new embellishments. For even though contemporary civilization is so largely urbanized that direct contact with him is, at best, minimal, the hog still looms as a substantial form that sticks to the mind-ribs.

The modern-day problem is that a large part of what people presently believe about him bears less real relation to the actualities of hogritude than has been so at any prior time in history. The view of the hog that is generally held today is, itself, myth. As urban man and rural hog have grown increasingly estranged in this fragmented society, people have tended to succumb more and more to the fictional pig pictures in their minds, and to their own prejudicial platitudes.

Some of the mythical spillover from the past is decidely pro-swine, such as the Chinese Year of the Pig (the last one being 1971, or Chinese calendar year 4669), which comes in a cycle of once every twelve years and represents a time for prosperity and material abundance. But much of Western citified hoglore has devolved into a cornucopia of clichés derived from garbled fables, nursery rhymes and senselessly repeated hog expressions or swine similes ("fat as . . . dirty as . . . lazy as . . . etc."). Even piggy banks, and the greed they imply, serve to confirm this discredited old hog consciousness in the minds of impressionable youth.

Perhaps the most dastardly disservice to hogdom is perpetrated by the urban American's reliance upon snappy images as a surrogate for thought. A steam locomotive, all smudgy and ungraceful, is termed a hog; a big, wide American-made auto is a "gas hog," while its driver is very likely a "road hog"; a stripped Harley-Davidson 74 motorcycle is a "chopped hog" (because in stock condition the machine is laden with heavy fenders and metallic gear and is hence "hoggish"); and "Lawrence of Poland," so the story goes, is identifiable from the fact that he wears a dirty white sheet and rides a pig. The unkindest imagery of all, though, is found in present political usage, wherein the ultimate condemnation for a public figure is "fascist pig" or "Nazi pig." Moreover, policemen in general are tagged "pigs," and males who display any lack of sensitivity or unwillingness to acquiesce to feminist demands are "male chauvinist pigs."

But elsewhere, fortunately, hoglore hasn't fallen to such a low state. Throughout the byways and bayous and backroad possum crossings of rural America, the convergence of country magic, backwoods wisdom and barnyard observations has continued to foster fresh sweet reams of porcine superstitions: like the belief that hog tails twist clockwise in the South and counterclockwise in the North; like the specific request in restaurants for a left-sided ham sandwich (since hogs are said to scratch with their left hind leg, thereby making the left ham more muscular). It's in the hinterland, too, that certain elements of ancient high hog ritualism are practiced—principally in the language and chants and prescribed way people conduct themselves at hog auctions.

In the course of our two species' historic continuity upon the earth, the paths of man and hog continually intersect. In myth and legend the two have come together to produce at moments a truly electrical interchange. Belief in hogs has historically altered human behavior. Yet in spite of all these interrelationships and interenhancements, in spite of all his intricate windings about our lives and the constant mirroring he offers us of ourselves, established society is still afraid to allow the hog to interact with us on a more personal and direct dimension. In the world of the collective mind, hogs remain bearers of omens, agents of foreign powers or ambassadors from some strange subterranean netherrealm we will never visit. Hogs have been used in so many symbolic ways that they have naturally assumed a whole aurora borealis of emotional colorations, which tends to make people nervous and uneasy about them. But even beyond that, we are taught, by implication and innuendo, to feel enormous discomfort upon perceiving any points of kinship in actual life. Any attempts at closeness or communion or any efforts honestly to comprehend the living creature himself are socially discouraged. Thus, hogs must be viewed on the level of mythical figures in order to become somehow comprehensible to us.

And what, you may ask, is the function of all these mythic-mystic-ritualistic interconnections? And what is the fundamental porcine meaning that arises from it all? The answer is that in the total sum of lore and legendry the hog signifies, in essence, the sheer miracle and ecstasy of naked existence in persistently reborn form. Just that: the fact that he lives and breathes and endures and extends himself in endless, inexhaustible reduplications.

And as such—as an ancient and unquenchably procreative creature always on the move—the hog brings his full weight to bear upon the everlasting human conflict between tradition and change. In this vein, we distrust him (and are awed by him) simply because we cannot possibly control or curtail him.

We can, however, control the way we want to think of the hog by means of projecting all manner of arbitrary traits and prejudicial notions upon him. "Projection" is the device of falsely attributing to others those motives or characteristics that are our own, or that in some way justify or explain our own. Maybe the fact that hogs are victims of our projection helps explain why human passions, angers and frailties seem most readily transferrable to them. On the Day of Atonement in ancient Hebrew ritual a live goat was selected, upon whose head the high priest laid both hands and confessed over it the iniquities of the children of Israel. The sins of the populace were thereby symbolically transferred, after which the people felt purged and, for the time being, guiltless. The goat was then taken into the wilderness and let go. This was the origin of the term "scapegoat." Of course there persists even today this same notion that guilt and misfortune can be shifted. But today the scapegoat for most things is the hog. On an intellectual level, to be sure, people may (and do) know better, and some even feel bad about it, but we continue to rationalize and make excuses daily.

The crucial and self-evident aspect of all this, which people practically bend over backward to avoid acknowledging, is that we are part of the same spacious tapestry, part of each other. Even on a physical level, we are, at least in part, the composite of all the hogs we and our ancestors have eaten, all the porkers who've gone to their final resting place, their graves, inside us. It is certainly understandable that this might be hard to swallow all at once, for the evolution of human awareness is not a smooth, steady continuum. Rather it proceeds in sudden discontinuous steps, in which each move to a higher plane of awareness is achieved by revelation, by a new view of the human individual in his wholeness. The concept of hogritude is one of these revelatory steps in the evolution of consciousness, and its acceptance would undeniably represent a real advance in our grasp of human possibilities. Of course, this is just a step along the way, for man cannot be said to have evolved in the ultimate sense unless and until there no longer exists within him a willingness to kill other things.

The question remains, though: What possibly dwells within us that still compels us to ascribe all our most base traits to hogs? Or can it truly be that hogs consciously absorb somehow the onus of our shortcomings as part of some celestial scheme? Deep down we know that to exist is to be related. Perhaps in this regard they have been assigned (or assigned themselves) to us *en masse* as bristly *Doppelgängers* . . . or, collectively, as a soft wall separating us brother beasts from certain harsher worldly realities we couldn't yet begin to comprehend. Perhaps they perform a psychological function for us as mythical entities in available and easily-abusable form.

Or possibly—with all of mankind's mystical, spiritual and philosophical gropings notwithstanding—if it can now be considered true, as some suggest, that God is indeed silent, then perhaps the hog is simply godlike in his inability thus far to make more direct contact with creatures such as ourselves. In which case, perhaps he may well have been put upon this earth for some glisteningly high holy purpose not yet manifest. Or, to speculate upon it from another point of view: The hog may be onto something. Or, once again, as Yeats mused:

> *Surely some revelation is at hand . . .*
> *And what rough beast, its hour come round at last,*
> *Slouches towards Bethlehem to be born?*

IX

The New Hog Consciousness

". . . And a little hog shall lead them."

> I like pigs and I honestly believe that most pigs like me. Hogs
> are beautiful. Some of my best friends are hogs. Like Patrick
> Henry, I care not what unfortunate remarks others may make,
> but as for me, I like pigs. Against all who would besmirch them,
> I stand ready to speak in their defense—even on the floor of the
> House of Representatives of the U. S. Congress.
> —Representative Fred Schwengel (R-Iowa)

> See those hogs? No man should be allowed to be President who
> does not understand hogs, or hasn't been around a manure pile.
> —Harry S Truman

And so now . . . now at last . . . now that you've been prepared for
it up to this point and are presumably "ready" . . . it seems safe to speculate aloud, with minimal fear of contradiction, that regardless of whatever their lowly status may have been in the recent past, hoglots—the
last and least glamorous American frontier—will henceforth provide and
presage the scenario for the mass upheaval and eventual change in
human society!

The very imminent plausibility of this notion already shimmers and
beckons before us. The potential, of course, has long been within reach—
much in the same way as hogs themselves have for so long now been patiently waiting in the wings for the proper cue to take their rightful place
upon the stage of earthly affairs. For what it all ultimately gets around
to is this: We—and by "we" I mean hogs and ourselves—are jointly abiding here side by side on this planet awaiting some hopeful opportunity to
transcend ourselves. And it is precisely toward this end today that the

hog willingly holds himself forth as an ideal agent for inducing us to break our narrow containments . . . to sever the bonds . . . shed our skins . . . clear the hurdles . . . and thereby scale new heights of enlightenment and psychic liberation. Clearly it is the hog—as man's best *real* friend—who most fully embodies our own hopes for the benign transformation of the world around us. In short, hogs can help smooth our transition into the future.

(I didn't want you to think this whole thing wasn't leading somewhere. The magic moment has arrived for you to begin reading very slowly and pay attention, for you are now, as never before, in a position to respond to what follows with a certain informed wisdom and uncluttered vision. Now that we have scraped off so many of the old barnacles we are at last able to sail into fresh waters.)

Dear Brothers and Sisters in the Cause (for you may now be properly addressed as such), by this stage of our odyssey it should be indisputably acknowledged among us that the all-pervasive essence of Hog has resonated across time and insinuated itself deep into the subtle convolutions of our collective mind. Throughout history, as we have seen, the common porker has been more than adequately patronized, lionized, mythologized and utilized—though even so he remains divided from our day-to-day selves by the glacial reaches of icy estrangement.

Essentially, the problem has been one of psychic distance. As things presently stand, the hog hovers in a sort of half-assed abstract way just beyond the boundaries of the average man's personal circle of psychological exclusivity. Nevertheless, he continues to touch and tug upon humankind in all sorts of cerebral ways. No other two species, for instance, continue to share such a natural subliminal rapport nor such an intricately balanced fascination-and-repulsion as do men and hogs. Moreover, there is no other earthly life-form toward which men so eagerly apply so many overgeneralized negative judgments. Hogs serve as catch-alls for the outcast passions of the human heart. They are blatantly slandered, mishandled and thought ill of NOT so much on the basis of their natural habits as *hogs* as on the basis of their shortcomings as *people*. Note that. The actual (unspoken) rationale behind man's hog prejudice doesn't have to do with the hog's status as a singular animal who may happen to possess this or that unsavory trait; rather it is simply that he is Not-People. And being Not-People he therefore merits no rights, no respectability, no individuality, no identity. He and his kin-swine are raised merely as commodities, as products to be harvested; under no circumstances are they permitted to be regarded with an eye toward personal contact or communication.

Nevertheless, our two species have contrived to thrive away together in relatively close conjunction through the ages without ever so much as once resorting to overt warfare. It's all part of a conscious pattern. As is true of other societal situations wherein beings of different shapes or races are thrust together, a *modus vivendi* is worked out among all parties. "A rutual of relations is established, and most people abide unthinkingly by the realities of the social structure. Since they merely follow the folkways they deny that they are prejudiced."[1]

The present relationship (if you can call it that) between our species amounts, in effect, to a form of corrosive collusion at the everlasting expense of the porcine race. And it goes on to such a thoroughgoing degree that each and every hog, for all practical purposes, is spiritually strangled at birth by the leechy tentacles of the System. Under the current regime, the hog is humiliated, maligned and all-around thwarted to such an extent that in spite of his god-given monumental fecundity he is rendered perpetually impotent. He is left all alone out there in his lot, befriended only by St. Jude, the patron of lost causes.

For when you get right down to it, hogs are victims of Technology, of technique. The techniques which undergird all our institutions and endeavors are, says Jaques Ellul, assimilating all of us into "a society of objects, run by objects." The ultimate effect of all which is that the means become the ends . . . and the original meanings of things become lost altogether. "What would we have to think of hogs," wrote James Agee, "who, having managed to secure justice among themselves, still and continuously and without the undertow of a thought to the contrary exploited every other creature and material of the planet, and who wore in their eyes, perfectly undisturbed by any second consideration, the high and holy light of science or religion." We remember the details; the whole escapes us. And so we go on about the routine exploitation of our hogs in the name of Agriculture or Industry & Commerce or Better Pork; and in the end it all contributes to the vast-scale devaluation of life itself, for one cannot deny the legitimacy of another creature without diminishing one's own.

Still, there's some cause for hope, of a sort. For one thing, there's a certain bright hopefulness to be found in the fact that everything you've ever been taught is largely wrong. And secondly, in spite of all we have wrought upon the porcine race, the subtle subcutaneous linkages between man and hog remain intact. Since it so happens that human beings are forever incomplete, we tend to look to the hog for fulfillment. If we were not incomplete without the hog we would not be desiring so many things. We desire because we are incomplete. We rely upon the hog in

many ways for support and for a sense of definition—definition of ourselves, for instance, as presumably superior, handsomer and all-around
more legitimate creatures. It's in this way that we subconsciously employ
the hog to help render existence endurable.

And yet even in the most ardent throes of our prejudicial passions
there do occasionally burst these momentary mental epiphanies, these
apprehensions of vague intimations and swift phantomic glitterings,
which strongly suggest that our current conception of what hogs are all
about bears roughly the same dim relationship to True Hog as the shadows on the walls of Plato's cave bear to Reality.

Leaving aside the aberrant fashionable attitudes of the moment, the
stark truth remains that due to centuries of physical proximity and spiritual interchangeability, human beings are—whether anyone likes it or not
—habituated to hogs . . . subtly addicted to them. Hooked. It's as if they
have been consigned by cosmic forces to fill the void at the center of the
human psyche.

The actuality of this condition is only just now seeping slowly into
our conscious minds. Ever so gradually, people are coming to take note
of that inchoate and disturbing beatitude, often mixed with distressing
hallucinations of happiness, that flows over them whenever they happen
to stumble into the presence of a hog. What it is, you see, is that they are
simultaneously discovering and reveling in that aura of serenity and eternalness he radiates. People are realizing, as one newly awakened young
Manhattan executive put it, "You can be your*self* with a hog!" Some are
even attaining that level of sensitivity and awareness wherein they are
able to regard large congregations of hogs no longer as merely so much
massed pork, but rather as collections of individual creatures, each
unique, unprecedented and unrepeatable. For in truth, no one ever sees
the same hog twice. The implications of this changing-of-heart are truly
awesome to contemplate. "Everything that people can't understand and
don't see the reason of," wrote Mark Twain, "does good if you only hold
on and give it a fair shake; Providence don't fire no blank cartridges,
boys."

And thus it is with the hogs among us who, in the eyes of growing
multitudes of pig-aware people today, are finally coming forth as full-
fledged spiritual dynamos generating epiphaneous and radiant little
glimpses of themselves which flicker bright and brief as summer lightning on the hills beyond the trees. In point of actual fact, there appears
to exist a sort of "social synapse" between man and swine as we thrive
away together here on the earth. Distinct signals are transmitted across
that bio-electrical juncture almost as if we were partners in some com-

mon and all-encompassing nervous system. I submit, therefore, that we are indeed inextricably joined . . . and not only that, but that we are poised right now on the lofty, rapturous brink of something hopeful and infinite. We as a species have just been suffering a momentary lapse until we get back in tune with the universal scheme of things.

For most, of course, the hog is still more admired in the form of refrigerated meat than as a bona fide animal or, beyond that, as a potential shaper of mass consciousness. But be that as it may, the actuality remains that just a few millimeters beyond the perimeters of current conventional perception there seethes a whole swine subculture, a separate reality, yearning to slip the nets and snares of ordinary existence . . . yet all the while living out their gentle, profoundly subversive lives as they bide their time as best they can, waiting until the day of our eventual breakthrough.

Although the hog may continue for the time being to rank in most minds as Not-People, nonetheless we do need him; not solely for his fleshly presence *per se* but for the momentous disclosures and spiritual revelations that only he is able to impart—serving, in effect, as an ax for the frozen seas within us.

Hence has it come at last to this: At a time of great travail (now) when the prevailing social fabric is worn gauze-thin . . . in this anguished epoch of deferred dreams, deteriorating passions . . . old values that dissipate bit by bit with each tick of the clock . . . in this era of squeamish allegiances and a diminishing reverence for life itself, we each find ourselves beset with muted *Angst* awaiting even a tentative blueprint for some new kind of heaven and new earth. Because the hog is a proven catalyst, because he stands forever on call as a spiritual catharsis and, more than anything else, because he—entirely within himself—hints of whole new dimensions of mind, we therefore must stand open to the onset of some titanic shock therapy, if that's what it must take, to escalate the reality of hogritude (and all the hopes contained therein) to the eye-level of general awareness. And here it is:

THE GREAT PORK REVOLT

◆

At length the everlasting quest had led him to this point. As dusk commenced to lower its obliterating pall, the young reporter, lone and weary, cast his eye upward, scanning and scaling the forbidding reaches of this one seemingly random mountain that rose up with no special distinction among all the other glum hills around it. Acting upon information painstakingly pieced together from a plethora of tips and confidential sources, he now found himself here—high in the gloomful strip-mined-out Appalachian ranges of Eastern Kentucky—extending once more his relentless search for the one living oracle allegedly able not only to make *sense* of what had happened but also to offer clues in unraveling the secret of the hog . . . and then, possibly, to put the whole business into some comprehensible perspective at long last.

In the muted hues of early evening the steep mountainscape seemed to grow ever more ravaged-looking and grim as he groped his way toward the top, toward the row of round mine shafts near the crest reamed out years back by the huge mechanical auger-drills of strip-miners—and now, sucked dry of coal, left like forgotten wounds in the mountainside, or like a vertical line of empty graves.

As he neared the crest, the atmosphere of creeping otherworldliness now seemed to spread at a faster pace: the terrain became even more leprously pocked and pale, and wrenched trees sprouted sidewise out of the cliffs. It was land no one would have imagined capable of supporting life. Yet it was here (from all reports, at least) that one of the last remaining leaders had been lurking in perfect furtiveness for some years now, making his den far down inside one of these augered-out abandoned tunnels, and disseminating from that point occasional pronouncements and bits of propaganda, if only to assure his fellow radical porkers across the land that he lives!

In the darkling air, the young man was able finally to pinpoint the precise tunnel by means of the ever-so-faint phosphorescent glow that emanated from within. And now, clenching his teeth in adrenalized expectation, he entered and inched his way far into the bowels of the earth. Deep. Deeper. Then suddenly, around a turn, the tunnel opened out wider into a sort of anteroom . . . and beyond that into a larger excavation whose walls dimly radiated a mineral luminescence revealing signs of animal life. A faint porkish tinge hung in the smuggy air. Here, too, from what he could see, were mats of pinestraw, a few crude implements and—aha!—there, right in the center, flanked by three compatriots, stood the very one whose sheer presence and stately deportment set him clearly apart from the rest: Alcibiades . . . Alcibiades the Prairie Avenger, elder sage and psalmist of the swine gospel, porcine revolutionary *par excellence,* preeminent figure high in the pantheon of hog heroes; branded by some an "agent of Peking," yet disclosing in his expression great reservoirs of gentleness and creature compassion.

No sooner had the young reporter stumbled abruptly upon this cryptic den of hogs than a stout boar stole from the shadows behind him and pointed a razor-sharp tusk to his backside, upon which he instinctively raised his hands above his head.

"Nay, Luke," said Alcibiades in a strong but aged voice, as he lumbered forward a step or two, "I sense he means us no ill, lest he would not have come alone." And at this the boar withdrew his tusk tip from the young man's rear. "Still," Alcibiades nodded, looking over his shoulder to the rest, "it gives us cause to remember once again that 'Eternal vigilance is the price we all must pay to keep something we don't like from getting at us.'" And at this the other hogs grunted agreement while their leader paused to burst into a brief spate of coughing.

Standing here in this shadowy glow, Alcibiades wore the sallow and lined visage of a senior swine. He was a little pale now, to be sure, and getting a bit long in the tooth, yet he peered forth with bright brown eyes that held the glowering incandescence of incalculable agelessness. Around his neck hung a pork medallion evidently awarded for his renowned bravery and crusading spirit; for here, indeed, was a hog with a palpable sense of mission, one who had long since left his trotterprint upon history. Though his body appeared to be slipping into bulky wrinkle and gauntness, he nevertheless strode fiercely about his lair with an undimmed animal energy, wearing a gray cloak of bird-down that flapped behind him as he moved, occasionally shedding a vagrant feather. For a few moments more, seemingly heedless of his visitor, he flailed restlessly about the cavern on the edge of some kind of rage, as the other three hogs, all silent, stood back respectfully, giving him room. Then, just as erratically, he fell into a fit of rhetorical despair. "Ah, what has become of us? What has become of us?" he wailed. "We are few of days and full of sorrow, forced to skulk and huddle here in this hole like fugitives, like moles, while out there in the land there remains so much left yet to be done . . . so much . . . so very much . . . woh!"

The great Prairie Avenger drew up still into a momentary silence; then, remembering himself, he turned to the visitor. "You are most welcome, sir, to share our trough and tarry with us for a while," he declared with great expansiveness. "I am . . ."

"Yes, yes, I know who you are," the young reporter blurted, sensing an enormous lifting of tension. The other hogs felt more at ease as well, and each now came forth for introductions. The first was called Porkheap of Truthful Tongue, a Yorkshire barrow, dry-witted and spry, though showing certain signs of advancing age. Porkheap duly bowed with proper pomp, and then, on stepping backward, flashed a quick wink as if again to reassure the young reporter that all was well with his coming here. Next, marching briskly around in front of the visitor whom he'd so recently held at bay, came the crusty old boar known as Cool Ham Luke, the All-American Hog, with the face of a Serbo-Croatian anarchist, a score of battle scars and a wooden left foreleg. Finally there was Frieda, faintly wraithlike for all her 400-plus pounds, who discreetly let it be known at the outset that though she had brought many a brood into the world she preferred to be known not as a dowdy *sow*, an abusive man-made term, but rather as a sprightly "swinette." She concluded with a sort of curtsy and tail-wag before retiring meekly to the rear to tend a pot of gruel she'd been stewing.

The three male hogs and the young man sat themselves beside the part of the wall that glowed brightest in order that he might take notes, and Alcibiades began. "We've been expecting you, you know. Or at least expecting someone . . . or . . . or some *thing*."

"No, I didn't know that," the visitor replied.

"Yes, it was just a matter of time," the elder hog proclaimed. "We are so well hidden in these hills that only the Lord knows where to find us." (And at this mention the other two instinctively crossed themselves.) "So clearly it was the Lord who sent you here."

"Well, not really, I . . ."

"Yea and verily, I say that it was," Alcibiades avowed in a sepulchral tone, with a touch of the prophet about him. "And now you want to find from us how goes the movement and to learn of our plight. And in return, *we* want to hear some word of how fare our brothers out yonder in the course of their ceaseless struggle."

"Ceaseless, just and *holy* struggle!" interjected Cool Ham Luke in tones of true believer defiance.

"Well," shrugged Alcibiades to the visitor, speaking slowly and resignedly in his ancient gruff voice, "what can I possibly tell you? *How* can I tell you? . . ."

"You can't! You can't even begin to *hope* to make him understand," exploded Luke, who now began to pace and gesticulate. "To grasp what we're about you first have to comprehend . . . to feel . . . to share the *hog experience!* You've got to sense the thwart, the pain, to eat the food, to taste the mud, to watch your loved ones parceled out to other boars like floozies at a poker party. And then you've got to live in certain dread of spotting that final hateful glint of blade. I've done time in a sty. I know. I *know*." By this stage the pitch of Luke's voice was climbing half an octave with each word as he lapsed more and more into a pat litany of revolutionary rhetoric. After carrying on in this vein a few moments more, however, the hot-blooded porker caught himself, then slumped back down and apologized to the group. Alcibiades began again:

"Aye, but there are things that even none of us knows for sure. It is to laugh, sadly," he said with a half-smile and sigh. "We grunt and fret and theorize, yet there linger questions no swine can answer for once and all. Like what does it *mean* to be a hog. But more than that, what *could* it, *should* it mean?"

Alcibiades halted suddenly to cough, at which time the ever patient Frieda, who'd been waiting just outside the circle, set before him a bowl of gruel which, though fresh-made, seemed frighteningly thin. "Alas," mourned the wise one, gazing into his bowl, "we've fallen upon misbegotten times. It has become such that just last week—and I am ashamed to say it—I caught . . . and ate . . . a robin who'd flown deep into the tunnel. Here I . . . *I*, longstanding champion of the concept of creature dignity, being reduced to a clumsy pursuer of birds and small rodents." Alcibiades bowed his snout in remorse, whereupon sweet Frieda promptly sought to soothe him by way of a few tender roots against his jowl. She consoled him some moments more with comforting murmurations in his ear and softly called him by the little names he knew before the great dream took him and he grew to be so mighty.

Now he raised his head, renewed. "Long ago it was . . . you'll have to follow me closely here for it sorely plagues me to tell, yet it must be known by all lest our lives here have come to naught."

"Amen, brother hog," said Porkheap.

"Tell it, tell it," Cool Ham Luke said.

"Aye, I will tell it softly," nodded Alcibiades, as he gradually began a mindless to-and-fro rocking on his hams and as his eyes smoked over with remote remembrances of that saga which has long since passed into porcine folklore. "Bear with me, trusted comrades, and correct me when I falter, for it's been so many, many years ago—thirty, maybe more. Ah, we were such callow young spry swine back then, heedless rooters grubbing in the hoglot mud, all so filled with lifely juices. And yet . . . and yet so low-rated . . . humiliated . . ." Here his gaunt face took on a forbidding cast as he found himself inexorably locked into the grim memory of that time, his dark eyes fixed upon some wild internal vision. Words long dammed up spilt forth in a relentless thoughtstream, punctuated now and again by factual corrections and occasional disputes over conflicting recollections by the Prairie Avenger's two lieutenants. What follows, then, is the version that finally emerged:

The countryside on that day back in the fall of 1947 reveled and basked benignly under a golden late afternoon of comfort and protection —everywhere, that is, except within the thousands of hoglots across the land, barren joyless plots which were designed (or so it seemed) to offer fewer of life's amenities than any other circumstances foisted upon any other of God's creatures anywhere else on earth. In retrospect, observed Alcibiades, every sty, pen, piggery and hoglot was each a separately contrived scene of deliberate degradation wherein languished groupings of perfectly honorable pigs persistently plugging away, payin' their dues . . . payin' their dues. Everywhere, high and low, they lived ensconced in this kind of uttermost dismality, waiting hopelessly for their future. It was hogness spindled, folded and mutilated on a monumental scale, with each porker compelled by precedent to blend into dingy conformity, each one destined to be ground down like his father before him, and in the meanwhile constrained to accept discomfort, grief, utter tormentation and malaise of mind all as part of the natural Sisyphus-like pathos of the pig condition—holding firmly all the while to laws that are not of the hog world but are actually directed against it.

"We were so *dumb* back then," Alcibiades winced. And here his two lieutenants snorted as if to say "A-men"; for they, like all other hogs, had similarly submitted in those days to this senseless subjugation, purely on account of not knowing that the business of being alive could possibly offer anything else. In essence, this, then, was life erected on the narrowest possible base: confined, cut off, disparaged, derided, barred from social intercourse with other creatures—an existence entangled in webs of loneliness and bottomless travail, all tension and menace, vaguely brightened only by the darkful fact that death came at an early age. Thus it was that each pig, as he ambled about in his indignity, amounted to a massive myth-shrouded hulk of potential repressed, passion compressed and spirituality ignored or held in suppression: raw energy exploding in a vacuum.

In those days, Alcibiades—incarcerated, as it happened, just outside Humboldt, Kansas—was popularly regarded from all outward appearances as a dapper young Duroc swain. Even so, every one of his trough-mates was equally well aware that there was far, far more to this young hog than met the eye. Though unlettered, he was not only precocious to an extraordinary degree but also endowed *spiritually* as a True Dreamer, a prophet, psychic, seer, a mystic visionary stargazer—and bound for market three brief weeks hence.

Late in this particular autumn afternoon, the more than seven hundred porkers confined here had been routinely "slopped" (a popular term of the time), after which most had gone back to wandering across the cataclysmic ruin of their lot, amid the standard cacophony of soft snorts and surly grunts. (The farmer, for his part, having executed his dismal chores with the required amount of insensitivity, repaired to his manse, where he sat slouched at the kitchen table getting high off bad spaghetti.)

Suddenly, across the leafy treetops, there coursed a strange zephyr. And at once Alcibiades, being the first to sense something afoot, or a-wing, immediately ceased his private rootings and raised his head. Came next a rush of wind that shook the landscape, followed by still greater gusts, all infuriate and scarifying—or, in any case, sufficient to attract the attention of even the most preoccupied or dull-witted rooters. Promptly all the pigs cast their eyes and ears to the sky, since clearly something extraterrestrial was taking place. And then . . . a gruffling sound, like unto an immense clearing of throat. Then a voice . . . a great voice as of a trumpet, a voice which proceeded thereupon to speak straight into their hearts:

"AHHHHHHHHHH . . . HEAR ME, HEAR ME, BOARS AND SOWS AND SHOATS AND BARROWS. YEA, HEED MY WORDS YE GILTS AND PIGS AND TINY PIGLETS. FOR I AM THE LORD, THE ALL-SEEING HUSBANDMAN . . . THE ALMIGHTY HOG OF HOGS . . . THE GIVER OF LIGHT . . . THE STRIKER OF SHACKLES . . . THE RENDER OF FENCES AND BARRIERS, VARIOUS WALLS AND VEILS OF TEMPLES . . . THE MELDER OF LIVES AND SPEAKER IN METER. AND NOW, AT LAST, I COME TO SET YOU FREE."

Some of the hogs were sore afraid but others, most in fact, were patiently attentive and contemplative. Still others were merely suspicious. But all, of course, were courteous. And the Voice boomed on:

"—I COME TODAY TO HELP THEE SLIP THE SURLY BONDS. FOR I HAVE KNOWN THY WOES . . . FELT THY PAINS . . . AND SEEN THY PROMISE STIFLED. VERILY I AM PLEDGED TO SEEK FOR THEE THY FULL FRUITION . . . FOR I DEAL IN IDEALS. THOUGH TRUE IT IS THAT ON A LARK I OFTEN TAKE A FANCY TO ACTS OF HUMONDEROUS VENGEANCE AND WHOLESALE MAYHEM WHEREUPON I SOMETIMES COME AS A SACKER OF CITIES . . . A PILLAR OF FIRE . . . A BURNER OF TENTS . . . DESTROYER OF HERDS AND FLOCKS AND FISHES. NEVERTHELESS I SHINE TODAY AS THINE OWN CHAMPION . . . THY ROD AND STAFF. AND THOU ART MY CHOICEST OF TEMPORAL CREATURES."

This was greeted throughout the lot by much hoof-stamping and oinks of exhilaration as many of the hogs looked at one another in amazement and then turned their heads back to the sky.

"THROUGH ME WILL THE BUD OF THY FULFILLMENT BLOOM. ALL BEINGS WILL LEARN TO QUEST FOR A ONENESS WITH HOGNESS. TO-WARD WHICH I FORESEE ONE DAY THY SMARMY HOGLOTS TURNED TO FIELDS OF FLOWERS. I SEE YOUNG PIGS AND CHILDREN RUNNING UPRIGHT HAND-IN-HOOF ACROSS THE GRASSY MEADOWS. BUT WOE AND BETIDES, IT IS HUMANKIND WHICH EVER SEEKS TO KEEP THEE FROM THY PURPOSE. AND HENCE IT IS I SAY: THIS DO YE, RISE UP . . . RISE UP FORTHWITH AGAINST THE DEMI-LORDS OF THE LAND . . . THE MAN . . . MISTER CHARLIE. GO YE TO THE FAR COUNTRY OF GLISTENING STONE AND GARDEN. STORM THE STOCKYARDS. BREAK THE BOUNDS. AND WREAK A VISITATION IN MEMORABLE PROPOR-TION TO THY MASSED MAJESTY, THY MANIFEST DESTINY . . ."

(By now the Great Voice, alternately soft and thunderous, held each and every swine transfixedly mesmerized. "No more the cruel knife," some thought they heard it say. "Restore the ancient bonds of family life," others deciphered. Still others swore they caught the promise of a "springtime for swinedom.")

Then, suddenly, with a subtle shift, the heaven-bellowed words lost connection altogether as the Voice slipped into garbled glossolalia—a fa-vored device of the Lord—which now, as He proceeded to speak in tongues, rendered him vastly more comprehensible than He'd been be-fore to some of the baffled porkers, who now heard the jumbled flow of sounds as a stupendous hogcall from out the sky . . . slowly fading northward, softening, diminishing . . . until, brief moments later, it came to be followed by a cosmical silence that stretched across the coun-tryside.

Here in the lot the hogs were stunned. But more than that: In the en-suing silence not a single porker lowered back his snout to root the feeble patch of earth beneath his feet, for each now looked out anew upon the bright inviting valleys and woodlands beyond. Plainly the Voice had given them pause. It was, as Alcibiades described it, the first dim ink-lings of what was to become a monumental hog awakening.

And life quickened. In the wake of their revelation, each single pig sensed something strange and unfamiliar within himself: the odd pulse of exuberation, the throb of certain hopes never previously permitted to fit into the context of his lowly life. The entire hoglot became infused with the heat and crackle of that protoplasmic kind of energy which makes a being want to stomp the ground and yowl down the moon.

Presently, an *ad hoc* delegation of responsible swine, along with a few hotheads, sought out Alcibiades for his counsel and sense of perspective. Other hogs collected nearby to hear. "My friends and fellow wallowers," Alcibiades began, "why we here have been singled out by higher powers I do not know. But I can declare for certain that from this point on our humble lives will never be the same again. The winds of change have come upon us. But the direction for that change is something we will probably have to choose for ourselves. Thus, our course must be decided soberly."

"But the choice has already been made," cried Baconofski, an irascible Poland China overmuch fond of fermented corn. "*We* are the chosen ones. There's nothing more we need to choose. Just act!"

"Ah," countered Alcibiades, "but the key consideration is where our actions may take us, or to what. Because thus far, as you well know, we've never even determined what it is we specifically seek. Now I myself feel our *first* aim must be to find ourselves a simple place away from here where nothing whatsoever is the same, a fertile site where hogs may safely graze. Once there, we can consider our purpose—if, indeed, we have one."

"Of course we have one," squealed Hamlet, another hothead. "Did not the Great Voice itself say that man wants to keep us from our purpose?" (And here the roused crowd of hogs grunted approvingly and nodded to one another.)

"True, true," replied Alcibiades, sensing restiveness among the herd, "but it failed to specify exactly what purpose we were being kept from." No sooner, however, did these words fall upon the group than there welled up a deep grumbling and the outburst of sporadic catcalls. Some hogs gnashed their teeth or pawed the ground in frustration, while here and there the words "coward" and "chicken" were passed about. And Alcibiades, hearing all this—and now increasingly fearful of the crowd's ire —injected a hasty, "But . . ." (and the other hogs hushed) ". . . but by all that's right and holy . . ." and here he fairly shouted, "WE SHALL FIND IT!" And at this the throbbing mob of porkers cheered, snorted and stomped approval. "*Man*," he continued, "*Man* knows what it is. And we're going to make him tell!"

And such a squealing went up—ahh—and grunts and grins and joyful oinks of jubilation. Some began to bugle and boogie. A clique of adolescent shoats on the crowd's edge broke into something of an impromptu quadrille. And then, at length, as the pandemonium gradually began to subside, there came up a plaintive urging from a couple of those closest by, a plea seized upon and repeated by a few more and presently by still others, and finally by the rest, all of them, all in unison: "Lead us," they cried, "lead us, great hog!" Upon which, Alcibiades, taken aback, slowly but gallantly lowered his head in humble consent.

Now the quick of the land stirred with hogs feeling their oats . . . hogs lost in thoughts of things wonderful and wicked, of oceans of corn and atavistic visions of truffles, not to speak of certain splendid notions never before even spoken aloud. But best of all, amid all this newly subversive stirring, the hogs here intuitively sensed distinct vibrations in the air indicating that their porcine compatriots elsewhere were similarly reveling in this strange eruption of spirit.

And indeed this was so. For, being welded into that common consciousness which binds swine together, hogs in lots throughout the land now simultaneously took an abrupt shift in stance, with the most determined among them going onto what amounted to full wartime footing. "Every hog a fighting hog," resounded as the common cry.

Toward this end, on the evening of that same day, there followed here in Kansas—and most everywhere else as well—a "speak bitterness meeting" wherein individual hogs recounted their maltreatment at the hands of their captors and stoked their choler up to creative peaks with fiery pledges to resist adversity and future subjugation. "We are the choicest ones," they crowed among themselves. "We shall enter magnificent cities," avowed some of the swine, while most of the others simply shouted back and forth: "Man seeks to keep us from our purpose" —still not entirely sure just what that meant.

And then . . . and then . . . right there in the full light of moon, a sleek black Hamp with a rich bass voice launched into an especially sonorous croon. His message touched a common nerve throughout the lot, for in practically no time his words were taken up by all his fellow porkers and escalated into an awesome sing-song hog-chant. In sorrow and exultancy they droned, as if borne of one single heart and voice: "LET OUR TUSHES GROW . . ."

The chant boomed and shimmered upward to a dizzying pinnacle of mystical intensity, then hovered there a moment, hovered, hovered and held itself in humming suspension as Alcibiades now raised high his foretrotter like a wand to give the solemn signal: And within that split instant the tempest was upon them . . . carrying them headlong into the dark land.

Thus did it begin. With a quick lurch of massed pig the barbed-wire fence was downed and out they burst, propelled in concert by the gawky grace of spontaneity and by their own unquiet blood. The great swine surge set forth at a timorous pace at first as the herd hoofed half-uneasily across the Spartan surroundings of hoglot litter, then past the barn and bits of farm equipment. It commenced as a slow, though stiffly joyous procession, rumbling, stumbling over moonlit fields, with the mass steadily falling into a certain rhythm that seemed laden more and more with menace—and yet, at the same time, with a great and exhilarating sense of liberation. It was a release of lifetime pent-up angers, aggravations, thwartations, hopes, heedless curiosity and no mere small amount of madness. And before long it picked up speed, bounding along at a gait which swelled and syncopated into the ominous harmony of mobilized hogdom on the move, moving ever northeasterly along the same general route the Great Voice had taken as it retreated beyond earshot.

In this manner—and with overtones of moral grandeur—the former confinees set their course and broke into a tireless run. They loped along undetected and pretty much in silence at the outset (most being understandably unaccustomed to their new status) until, some few miles distant from their old digs, a surprising thing took place. Suddenly, off to the left, there appeared a muzzy band of grunters on a hillside. Then more and more new pigs poured over the slope, running directly toward them. And soon, lo and behold, the original militants finally found themselves linked up with another whole herd, making them altogether more than 1,500 strong, with Alcibiades the Prairie Avenger trotting stoutly in the forefront. Here were hogs loosed upon the valleys and woodlands with an all-consuming delirium of spirit, shared, as if infectiously, by each fresh contingent that emerged and merged into the flowing wave.

Onward they plunged, tirelessly and apparently undiscovered, across the rural darkness, being enlarged upon every moment not merely in terms of their total number but also in each single hog's sense of slightly manic pixilation and the weight of earth cut free. It was a completely ecumenical movement encompassing hogs of all possible breeds, creeds and dispositions—and sunrise found it swollen into tens of thousands, still growing, still pounding across the countryside, now into Missouri. And so it went throughout the day and following night, with miscellaneous pigs along the way melding singly or in herds into the larger tide as it pressed forward, pausing in vacant fields or forests during the hotter parts of the afternoon to rest and be replenished by the fat of the land.

By the third sunrise the rushing horde of pork had crossed into Illinois, whereupon it joined forces with yet another vast sea of swine headed by a Hampshire known as Soldier, "Swiftest and Strongest Lead Hog East of the Mississippi." The two armies combined both bodies and battle cries as well as they charged fiercely forth now across Illinois chanting, "LET OUR TUSHES GROW" and "ONE HOG, ONE SOUL." Not too much longer thereafter, coming up south of Joliet, they happened to converge first with Cool Ham Luke's tough-faced porksters from New England and the industrial Northeast, and, next, with Fatback Beauregard's elite troop from Northern Virginia.

And on they roared like a divine wind, on, on toward Chicago: fierce inflexible legions thrust forward at last by the instincts of 40 million years—fever-seized, god-goaded, spirit-searing, unstoppable. The orgiastic thrumming rumble of their trotters reverberated through the countryside like a tocsin, or a call to insurrection. And everywhere their ranks were fattened by newly escaped herds as well as by hobo-hopping bands of wanderers or otherwise rootless rooters. Millions of them were there by this stage, and as they bore north toward the windy city their hoofbeats gradually slipped into a mystic cadence, a perfect pounding unity—a primal sound that seemed to resonate clear through the soil to the center of the planet itself, to the very quivers of the earth's core. It registered 8.4 on the Richter scale. And onward they swarmed across the early midday brightness in a fine clean surge: 16 million porcine knights with sun-bleached hair and eyes of morning-glories!

With their leaders in the vanguard setting the course, the hogs steered dead north for several miles more; then, in a wide arc, swung due east. The lowering sun hung behind them, just above their backs, as now the flurries of dust pressed out of the earth by their trotters pulsed high into the air like soft explosions of gold. And, lo, all at once, just over a rise—and across the dust and mist and miles of distance—upward loomed Chicago, with its haze-draped skyscrapers appearing to the hogs like soaring Olympian peaks amid blurred and far-off orchards. For many a porker the lacerating splendor of the moment gave rise to grand and swimming sensations of success, while for others it evoked the happy ring of pure hallucination. But nonetheless, all, to one degree or another, felt the pull of some deep undertow, enforced now by the inexorable push of their own immutable instincts. And now, in stupendous union, the pace of the huge hog armada picked up even more powerfully, with every single swine's instinctive and unappeasable hunger for contact being heightened more sharply than ever by this enormitude of richest expectation.

With their destination growing larger by the moment, the mind of each onrushing swine now began racing even faster than his galloping hooves. Individually they began nursing frantic visions or rehearsing parts in preparation for whatever it might turn out to be that was expected of them. Most weren't quite sure. A considerable number, for instance, seemed to recall, or else maybe just imagined, that they had come all this far with the aim of showing the world the sheer majesty of hogritude, as if "trooping the colors" in some grand form. Others, caught up inside the teeming, rolling throng, thought they'd all simply been seeking out some pleasant spot where they and their young hogs and their hogs' hogs might safely graze. Many sensed somehow they'd been called forth for undisclosed sacred purposes or to make certain sorts of impressive revelations yet to be announced. Some of the younger ones, in league with the more intensely discontented and those who were always eternally disgruntled, felt moved within their own hearts by a near-rabid lust for justice, or vengeance. While still others fancied that the overriding goal in all of this was to create a full-fledged groundswell of sufficient magnitude so as to bestir all fellow creatures to hop aboard the hog bandwagon. And lastly, a widely scattered few among the advancing porkers held dearly to the rumored notion that the entire lot of them was running to a predetermined spot to petition for a special dispensation from the Pope.

Nevertheless, all alike shared the unutterable thrill of being on the verge of some splendid new reunion with their primitive selves. And so . . . barreling out of the west with a billowy heave . . . the combined might of the swine millions gushed straight toward Chicago in a weighty tidal wave, which caused the very land itself to tremble:

. . . And elderly bankers and jaded portly gentlemen from Winnetka, riding their golf carts in flashy cowboy costumes, felt the distant tremors and frowned at one another gruffily in puzzlement;

. . . and well-scrubbed swingers playing paddle tennis in Des Plaines watched their aims go awry from this uninterpretable rumbling and said, "Hmmmmmm";

. . . and Oak Park housewives heard their china rattle or saw their laundry flutter on the line;

. . . and on the farthest edge of town, sleazy hoboes hunched in doorwells—seeing old cigar butts on the curbsides strangely start to roll—rose from their roosts and hobbled to the sidewalk . . . knelt, grabbed a butt, then straightened up, turned . . . and—*WHOOM!*—even before their brains had time to ice over at the paralyzing sight, the monstrous stone-crumbling onslaught was upon them: sixteen million porkers plunging madly out of the sun in an earsplitting swarming blur—hogs aflame, foam-flanked and ominous, tusks bared, ears flapping with shivering glee, their eyes opened all wide and triumphant.

Eastward over Riverside, through Berwyn, Cicero, on they poured in an all-crushing engulfment. And as they began streaking through the suburbs in thumping percolation they also began now to chant, chanting in perfect oneness, up and out from someplace desperately deep within each of them: "WE ARE THE CHOICEST ONES . . . WE SHALL

ENTER MAGNIFICENT CITIES . . . NO MORE THE CRUEL
KNIFE . . . LET OUR TUSHES GROWWWW . . ." But with the sun
orangely glaring from behind, hardly anyone who saw them could make
them out quite clearly, if at all; and not a single person in their path was
able to understand their chanted cries, which sounded for all the world

like senseless animal yowls and shrieks, all submerged into the cataclysmic staccato of their trotters . . . and further drowned out by the direful roar of shattering glass and tons of concrete crunching underfoot or shearing off the sides of skyscrapers.

At about this point, however, a goodly number of the pigs began to grow blindly frightened, finding themselves beset all at once by city sounds and sights they'd never experienced before, and lost themselves in sheer centrifugal animal fury, thereby pressing the vast hog mob forward faster and faster as looming shadows of strange stone buildings and unfamiliar urban shapes leapt and grimaced with unspeakable frenzy, crumbling down around them. To many people, stricken dumb by panic, it was like some ancestral amorphic beast descending, or perhaps the dawn of Armageddon; while those few citizens who were struck by the abrupt comprehension of what was actually taking place wandered helplessly off into mute ravings.

In the course of its charge, the hog juggernaut proceeded to lay waste to as much of Chicago and its environs as lay along the route of a wide path that lunged in a straight line toward the city's heart. Here and there, a man fell and was trounced flat by thousands of pounding trotters. Cement cracked, cars were smushed, lines were downed. An AMVETS parade (with military marching band, majorettes and a float featuring survivors of the Battle of Belleau Wood) took an inadvertent turn onto one of the boulevards awash with running hogs and was utterly routed and annihilated. Meanwhile from all parts of the city nervous sirens howled from the cars of policemen dashing about in all different directions, totally powerless to subdue what, by now, was officially designated a "Major Hog Riot."

The multi-million-pig phalanx cut cleanly across the dingy city, slicing onward until, as if by fate, there came into view the huge and loathsome Union Stockyards up ahead. In mere seconds, the hitherto hateful site was leveled, thereby releasing thousands more long-suffering kinswine who merged into the mass thrust—which, even more wildly than before, managed to crumble and rubbleize every edifice before it.

For *their* part, most of the more serious-minded hogs in the leading edge discovered suddenly that, like it or not, they were physically unable to stop at this stage, being pushed along by others in the rear who were running forward out of sheer hysteric dread lest they succumb to the mundane terrors of this urban ooze that surrounded them. Thus, the weaker ones kept the stronger ones going. But be that as it may, since no formal delegation of humans had come forth officially to treat with them there didn't appear any real reason for stopping now anyway. So, undimmed, they swept on heedless, harum-scarum, as, above them, brief tears splashed across the sun.

And presently, the winds seemed to pick up sharply, just as on that day the Great Voice had spoken. Maybe the Voice was about to speak once again and would tell them what they were supposed to do now. The winds gusted stronger as they ran on until soon a vast pellucid plain of some sort appeared to shimmer up ahead in the distance. And then more wind. Alcibiades, still running, raised his head to the sky—and promptly tumbled into an open manhole.

The winds by this time had accelerated to near-gale force. Surely now the Voice would speak. And now—with the innocence and immortality of their hopes newly recharged at this prospect—the rampaging animals thundered forward in glittering glee over the pavement . . . over the grass . . . over a stretch of sand. Here they charged full-tilt, with a long-last feeling of freedom; on, on, with a booming "LET OUR TUSHES GRR . . ."

. . . And at that split instant the wide plain they'd seen suddenly parted open, agape, yawning wide, wider—*AHHHHHHHHHHHHH*—and then, as if with a gigantic closing of great watery jaws, engulfed the entire headlong plummeting torrent of hogs: boar, sow, shoat, gilt, barrow, on, on and on. Each and all became folded softly down beneath a covering shroud of greenish-blue like so many babes in blankets. And none returned to the surface, not a one: the total mass extincted in a single wet gulp, all sinking down and down, sending up the desperate bubbles of their lungs' last air. With this concluded, the only sounds that rose to fill the echo of their absence now were soulful idle lappings at the shoreline punctuated with the low woan of ancient freighters farther out on Lake Michigan.

Picking themselves up from the aftermath of this devastation, the majority of traumatized Chicagoans freely acknowledged among themselves that the fearsome occurrence they'd just witnessed had indeed been due, somehow or other, to a strange and awful deluge of hogs. But over the next few weeks, disputes began to crop up concerning the actual cause of the whole thing as more and more people convinced themselves that it had really been the work of a tornado, or possibly even a vagrant shock wave from one of the flying saucers so newly popular during that period. Those few who persisted with the hog explanation were eventually persuaded that they'd been dazed insensible by falling debris or drunk or otherwise hallucinating in some fashion. What's more, since the entire activity transpired just prior to the video era it obviously did not become a televised event. And this being so, it developed years later that no one, not even those who were there, could say for absolute certain that it had ever happened to begin with. "If it didn't happen on the TV screen," declared one wag from the local Jaycees, "it most likely never took place at all. Probably *couldn't* have occurred. Any kid can tell you. It's as simple as that!"

Alcibiades gave a snort of pained disgust at this point in his recollections and lapsed into a brief bitter silence, terminated by another snort. "So *that's* what it finally came to," he grunted (interrupting himself now to cough violently a few times and then clear his throat). "An ending that must surely rank among the rages and failures of the universe." Porkheap and Luke looked equally downcast, but said nothing. "And now," continued Alcibiades, looking wan and elegantly despondent, "I . . . along with these few of my comrades who also managed to escape the fate of our kinswine . . . dwell here in the sole hope of keeping the hog flame alive. And I'm sure you realize this job has been bloody hard, too. Particularly what with the massive reprisals that went into effect right after Chicago. Losing all those brothers and sisters in that tragic engagement wrought all manner of havoc to the Cause. The immediate result was that the price of hogs and hog futures went up, which led, in turn, to more and more and more of the same. It was a backlash of hideous proportions—a virtual hog purge, a full-scale repression in the name of restoring 'reason' and 'the old ways' and 'more efficient swine management' that has continued even up to the present time. In fact, it so happens that even today hogs throughout the land are waxing and waning away at a greater rate than ever before. Even as we sit here. And still suffering . . . and still enduring the everyday death of spirit."

A gray tear rolled from Alcibiades' aged eye, and as it struck the ground he quickly cast his snout toward the ceiling and continued to emote with the doleful fervor of a martyr. "And here I . . . I the Prairie Avenger . . . am rendered helpless, forced to skulk beneath the earth like a felon left to root or die, while daily I grow older and adipose tissue collects in my thighs! Oh, God," he croaked, then coughed sharply.

The venerable old boar lowered his head. "Everything is yearning and dream," he reflected, "but when your stars are stacked you just can't seem to win." He paused, pondered . . . continued: "Aye, the outer world may pay little heed or mind to what our little band of weary firebrands do here. But it must never stop in its striving to comprehend the hog *experience,* nor cease to seek after the glories of true hogosity as it's practiced out in the provinces. For a hog, as you know, moves in many beautiful and mysterious ways."

Here Porkheap spoke up. "A hog is *joy!*" he proclaimed. "A hog is a way of being—a way of seeing. We are critters who wring ecstasy from the simplest glees and commonest of goals: a cottage on a quiet street . . . a faithful mate . . . a touch of love . . . a taste of respect . . . the liberty to stand in rain and let the water course down our bodies . . ." His voice trailed off as his eyes wistfully refocused somewhere far away. Then, with a start, he was jarred back to the issues of the moment by the unexpected entrance of Boris, a bushy-browed porker in a claw-hammer coat who had been out browsing for acorns all this time. Boris appeared a tad uneasy at the young reporter's presence until Alcibiades assured him, "It's all right, my lad, he is here to help us," and nodded sagely. On this note, Boris grew more relaxed. "Well, ring my snout!" he declared, then proceeded to unload acorns from his pockets.

Frieda, having set out lukewarm bowls of gruel for everyone, presently retired to a corner to plunk the zoad, a medieval hog instrument designed to be strummed softly by snout or trotter. Thin veils of semi-discernible music pulsed through the cave as Alcibiades bore on, the enormous energies and old appetites still clearly churning within him. "It was an exquisite moment, that mass arousal of Chicago," he thought back fondly, "a great gesture of freedom. Oh, I know we made mistakes. Some speculate that the Lord used us as agents of retribution for something or other and that we never should have listened. But ah," shrugged the great swine rebel, "who can say? Yet no matter what others might claim, it was still the first real breakthrough modern hogdom has known. And we just cannot let it be forgotten! It set loose in the land a fresh sense of swine pride and self-awareness. We found out we have *poetry* in us, by God! And lots more besides. Good feelings about life and stuff. Aye," he affirmed with a knowing squint of the eyes and a certain cryptic profundity, "hogdom will rise again. Mark my words. It *has* to. It's been such a long time a-coming, but *this* time we'll be prepared. It's just like some sagacious pig*frère* put it: 'To know how to free oneself is nothing; the arduous thing is to know what to do with one's freedom."

"Yes," responded the reporter, "that was Gide."

"An outstanding hog," sniffed Alcibiades, "and thanks to ideas like his the porcine race is going to reach its rightful place in the sun, next time around. What it all hinges upon is the question of who we are, you see? But soon we'll get around to a glimpse of the light at the end of the tunnel and make this a land fit for hogs—or at least inspire the powers-that-be to leave the door open for negotiations."

He sighed and glanced around the cave, gathering his world about him—and yet still distinctly presiding over it all like the prime minister of a government in exile. "Nay," he averred, "it hasn't ended yet. Nay indeed. For with us, you know, every defeat simply inspires a new hope. The way we see it, every moment of every day is part of the big countdown to our eventual fruition!" He announced this and, with a great foxy charm, halted briefly for effect, as his bottomless eyes flashed spacy twinkles. "The way we see it, the hog is simply waiting it out for now—waiting here, and in hoglots all over the countryside. Waiting, just like Judy Garland, for that next sweet chance to make another new comeback."

The magnificent old Prairie Avenger, plainly roused to revolutionary heights at this thought, now struck a pose like a Viking, drawing himself up proudly erect with his chest thrust out, his tattered cape draped regally across his back, snout flared, smiling warrior-like in the phosphorgreen glow.

"Well, yes, I . . . I understand," replied the young reporter with some hesitation, "but . . . well . . . Judy Garland's dead."

Alcibiades held frozen for a second. Two seconds. Three seconds. "Oh," he said in a meeker voice . . . and then began absentmindedly nosing in his bowl of gruel.

❁   ❁   ❁

In his book *When Will Civilization Stop Fooling Around and Get Serious for a Change?* Dr. Ranzbottom launches into a most eloquent discourse on the implications of the connotations of the possible cosmological ramifications of everyday porcine behavior and concludes by saying of hogs, "They've got their ways!" And their ways, more often than not, can best be made understandable to us by means that are more experiential than intellectual. They themselves instinctively know so many of the things we have to wade through reams of academic treatises to discover. Optimistic, genial, naïve and yet profound, noisily energetic and yet tenderly meditative, fertile and careless and potentially infinite, the hog appears to have been put upon the earth for some high and possibly holy purpose not yet evident to us.

As a consequence of his present neglect, the hog everlastingly carries within himself the intimations of an undefined mission in life. Merely to observe one or more pigs in their natural digs is to bear witness to the presence of some subsurface purposefulness of immeasurable intensity which they pursue with aimless and all-consuming passion. Above all else, and in the face of everything, hogs persist.

The fact is, the hog is an energy form that remolds and redefines himself continuously by his actions. He is both ends of any spectrum. In this regard, the hog is (as much as he can be said to be anything else) a state of mind—a tangible apparition, perhaps, of whatever you happen to fear or fancy. And thus it is that the single most vital issue the hog calls forth is whether you perceive him as a challenge, an insult, a comfort or a call to arms.

Meanwhile, our general awareness of hog—like our actual exposure to him—remains very effectively restricted. He is simply off-limits to our comprehension and compassion. Even on occasions when the average man chances to come upon a hog, what he sees is not the unvarnished creature in and of itself but a negative *description* of "hog" as determined by a consensus arrived at hundreds of years ago. And it is in this way that social custom, with the false attitude it fosters, serves to maintain an aura of tension which stifles communication and holds back the natural evolutionary extension of human awareness.

The average mind, if left more or less to wander about on its own devices, is naturally predisposed to soar beyond the boundaries of prevailing mental structures. Within the framework of current wisdom, the ultimate compelling force behind almost any conceivable point of view or plan of action is "objective reason." But today, more and more of the perceptions rising into our awareness are being discovered to be wholly outside the jurisdiction of standard rationality; and the upshot of trying to fit these new experiences into the old structures is that things just don't seem to jibe and fit quite right anymore.

Even now, the most obvious thrust of the cosmos is toward universality and an awareness of the oneness of all life. But this is a vision which is fundamentally at odds with intellectual reason, owing to the fact that pure intellect invariably separates and relates to phenomena in terms of their differences. Oneness is not "reasonable."

But in spite of it all, certain messages *do* manage to seep through the filters now and again. Like the idea of plant consciousness: the fairly recent but now widely accepted realization that plants show emotions, form attachments with other forms of life, possess memory and even display the ability to communicate.

Plant consciousness, though, seems somehow easier to accept than the idea that we have the same sort of extrasensory alliance with animals. (Indeed, if ever the question arises, the great preponderance of people today will concede that they themselves *are* animals, but you can tell that most of them don't really believe it.)

Nevertheless, the non-rational reality of plant consciousness is generally acknowledged today chiefly because acceptably scientific means were found for making the link between man and plant accessible and "visible" to contemporary minds.

This naturally raises heady hopes concerning our potential respect for life on the animal level, beginning, logically enough, with the hog. And actually, it just so happens that right at this moment we may very well be poised on the brink of the big breakthrough in this area as based upon the increasing evidence of the hog's visibility in the social mind's eye. More so than ever before, visions of hog continue to pop up before us in all manner of unexpected juxtapositions and fresh, consciousness-raising contexts.

The pig's present-day climb toward prominence first became apparent in the mid-1960s. Almost as if through some spontaneous eruption within the collective unconscious, events commenced to occur and situations began unfolding which, taken together, made hogs and hog-related concepts stand out in high new relief against the sodden back-drop of an otherwise hopeless civilization. All at once, people high and low began feeling something.

Hogs have become a *nouveauté sensationnelle* in all fields of creative expression—thereby demonstrating the general drift of the public mind. In the theater, for example, appeared the aforementioned play called *Futz!*—the tender and moving moral tale of a farmer who loved a sow, literally—which had such a stunning initial impact that, according to one reviewer, ". . . at some performances members of the audience have actually crawled out on their knees in a horrified effort to remove themselves from the deeply disturbing events onstage."

In the hugely diffuse realm of the counterculture, hogs are steadily acquiring all sorts of fresh significances. A communal group called the Hog Farm gained a certain notoriety in the late 1960s by traveling all over the United States in schoolbuses, along with their pet pig, performing multifarious acts of Samaritan decency and public service in the name of "hog consciousness." Then too, the Yippies and young radicals who converged on Chicago for the 1968 Democratic National Convention went so far as to nominate and present their own presidential candidate in the being of a young porker named Pigasus J. Pig, who went on to win a surprising number of votes in the national election.

It was also during this time that new hog heroes began insinuating themselves into the public's awareness. Notable among these was an underground comic book character created by cartoonist Gilbert Shelton and known as Wonder Warthog ("the hog of steel"). Though he is an ungainly sort of figure, WW is clearly a superheroic swine who, complete with cape and mask, roots out social evils and then reverts to his regular guise as a mild-mannered reporter.

Along with all of this, hogs and thoughts of hogs began simultaneously to blossom forth in virtually every medium of communication.

Suddenly—*WHOOSH!*—here were hogs on TV shows, in commercials, on billboards, posters, in poems, rock lyrics and letters to the editors, not to speak of the hundreds of thousands of "Hogs Are Beautiful" and "Pigs Are Pretty" buttons that began appearing on people's lapels.

Not all of these manifestations of public awareness have necessarily been favorable to hogs, it must be admitted. "Pig," for instance, gained a certain widespread currency as a pejorative term to be used with highest contempt against members of reactionary elements and various custodians of the *status quo* such as policemen, fascists, capitalists and male chauvinists. This is ironic inasmuch as pigs themselves would be the last of all creatures to condone the *status quo*. They, in fact, are at absolute cross-purposes with civilization in its seedier modern aspects and—if the truth were known—are actually anarchists. Nevertheless, the use of "pig" as a derogative is at least further testimony to the fact that society is thinking along porcine lines, and as the old politician puts it, "It don't make a damn what they say about you as long as they spell your name right."

In this context, a significant effort to clear the tarnished name of swinedom emerged in the recent proposal of Congressman Paul Findley (R-Ill.) to create a commemorative stamp honoring the hog. Hogs, he noted, had been pictured as symbols of pollution in a "Keep America Beautiful" campaign. "The bad image hogs get as 'villain of the environment' is enough to make us country boys weep," he said. "Fact of the matter is, the hog has made a tremendous contribution to man for more than nine thousand years . . . Surely he deserves the honor of a commemorative stamp. It will help show that hogs are really beautiful and to erase their undeserved image as dirty polluters."

These and other instances of the hog's new visibility are but small iceberg-tip outcroppings of the much larger cultural shift quietly taking place in men's minds. Swelling numbers of souls all over the landscape are not only starting to notice the hog for the first time but are also beginning to spot patterns and make connections they'd never before permitted themselves to perceive. And interestingly, the first effect of this new awareness within the individual, just as you'd expect, often turns out to be an initial period of remorse and repentance; for today, as in Eden, knowledge doth precede shame. But it also precedes change and positive psychic growth, particularly on the part of those individuals such as accept the truth gladly—those who are open to visions and are consequently not unwilling to reach out on a one-to-one basis to establish honest contact with a member of the porcine race.

Most city people, to be sure, are understandably hesitant to experience their inaugural confrontation if it must be in the hog's rural habitat. The chief justification for this hesitation is that the hog's own turf is known to be maintained in such a way as to repel potential visitors. This being so, it will probably prove necessary for those people concerned for the future liberation of society to seek out suitable porkers and bring them into their own homes for face-to-face encounter—and eventual breakthrough.

Those seriously seeking the road back to a more holistic vision of the world will probably discover that the best first step is the group encounter, involving both people and pigs—this being a practical opportunity for striking up that initial rapport so crucial toward man's formulating his future relationship to hog in new terms.

A historic encounter of this nature took place quite recently in Atlanta, Georgia, thusly thrusting a score of responsible citizens (including lawyers, business executives, a theater director, Joe Cumming from *Newsweek*, Reg Murphy, now president of the National Geographic Society, and George Leonard of the Esalen Institute) together with one hog, a delectable young Yorkshire gilt named Doreen.

At the outset of the evening, the human participants (most feeling somewhat awkward in their anticipations) arranged themselves in a circle on the floor of the living room, into the center of which was led the lone hog. Though Doreen was similarly apprehensive, she was also curious, and after a few perfunctory roots at the rug she explored the perimeter of the circle, diffidently nosing people's knees while receiving their timorous pattings and scratches. Suddenly, however, an opening appeared in the circle, through which Doreen escaped into the rest of the living room and other parts of the house. But then—in what became the tension-shattering moment of the evening and the first opening of doors to honest contact—she sauntered back into the center ring and gracefully sat herself down, confidently eyeing the members of the group. The hog had arrived!

Now she prowled about the circle peering deeply into everyone's eyes and being responded to, in turn, by each participant after his or her own fashion. In mere seconds, Doreen became the all-purpose repository for individual emotions. She became an abstraction to one member who fell straightaway into a frenzy of pig quotes from literature. One young lady (Jewish) rather clearly viewed the hog as a veritable labyrinth of forbidden sensations which caused her to flinch and hug herself in fear whenever Doreen approached. The director saw the porker as everyone else's "surrogate for the Negro," while others interpreted her as a creature to trust . . . or to mother . . . or to welcome . . . or pity—or as a source of certain formless suspicions. Interestingly, hardly anyone, at first, saw the hog as hog; rather, she became a deeper mirror, an ultimate projection, of each person's aspirations, or of whatever each may have disavowed within himself.

But in time, as the encounter progressed and people began loosening up to the point of being able to relate with greater empathy to that which was before them, these earlier perceptions seemed to fall away. Like some simultaneous dawning, a few individuals at first and soon virtually the entire group as one began looking at the hog with widened eyes and seeing her not as a reflection but as an equally legitimate part of the same elemental vibrancy, the same unfolding energy process in which they themselves were engaged by the simple fact of being alive.

The members of the group began rising above themselves as this realization deepened, and there was a fresh sense of joyfulness that floated almost palpably in the air, and laughter. People began hugging the hog with genuine endearment and a sensation of great weight cut loose. "To discover the truth in anything that is alien," Leonard Cohen advised, "first dispense with the indispensable in your own vision." And the hog provided the way. If "love of hog passeth human understanding," then forsake normal human understanding. Leap out of your skin and relate only in mystifications. Feel the grandeur and power of a radiant, slow dissolve into the air where it is possible to catch faint inklings of ourselves at last as kindred swine beneath the skin, brother porkers flowing together, equally inviolate . . . ahh!

Indeed, one of the major observations to unfold from this particular encounter turned out to be consistent with one of the historic themes recurring throughout hoglore: namely, that humans *find* themselves in the depths of the hog . . . that his elemental presence somehow accentuates the anxieties and absurdities we feel. In that regard, this encounter served also to demonstrate that on the eyeball-to-eyeball level you invariably verbalize or otherwise express to a hog exactly those things you honestly feel. After all, there's obviously no sense in *lying* to them.

As for its long-term effect, a hog encounter, theoretically speaking, cannot help but fill in some of the brain's blind spots and initiate a tentative rediscovery of our common bonds. It is a chance to confront the old ignoble defenses against feeling and, as such, is a vital step toward declassifying the hog so as to make it possible for still more people to wrap their minds around him. And it is by means of such revelations that people might then probe onward to discover—just as many have always sensed deep down in the shank of the soul—that hogs and humans are actually synergistic. For we learn to incorporate the hog as the hog incorporates us, each acknowledging the other as a legitimate fellow practitioner in the odd folly of being alive.

The history of science makes it clear that the greatest advances in man's understanding of the universe or of himself are made by intuitive leaps. There are all manner of things we have come to "know" that we don't yet understand in the quantifiable, logical sense. We do not fully comprehend electricity or magnetism or light, for instance. And we may *never* understand hogs, other than knowing that life does somehow feel different viewed from the inside of the soul of a swine—different, and yet very much the same. For the hog provides a new intuitive grasp of the fact that all things that are alive are brothers in the soil and in the sky, and you can believe it with your blood if not your eyes.

Way down inside somewhere, most of us accept the proposition that a civilization may be judged by the way it treats its animals—as this, for the most part, is a valid reflection and measure of its people's relative reverence for existence itself. The Hog Movement is rooted in just such a premise as this: in the view that every single act of living is a thing holy and worthy of awe; that all fleshly creatures are subtly interrelated segments of an infinite mosaic; and that within this one universe everything has a relationship to everything else, and also has a responsibility to everything else.

Early man's most innate cosmic orientation was toward a unity, a symbiotic interblending into oneness with the entire landscape, accepting everything—forces of nature, beasts, plants, fellow cavefolk, dreams, stones—all as part of a mutually shared identity. It so happens that hogs, highly sensitive to natural rhythms themselves, are uniquely capable today (or will be soon) of effecting a mass restoration of that old cosmic orientation and realization of oneness—a sense of oneness that extends to everything, even cows. Yes, even to cows. (More about the cow later on.)

The members of the Hog Movement comprise an astoundingly large but dispersed segment of the population, operating sometimes in small or moderate-sized groups and sometimes just in twos and threes. It's always a little jarring to come upon new evidences of the fact that there actually thrives a cult or whole subculture of people from all walks of life who honestly *care* about hogs or are intrigued by them for one reason or another and who, in general, believe that society should somehow consider effecting a new social contract with the creatures. The collected sum of their separate attitudes and most recent realizations about the hog is a fairly comprehensive statement of possibilities that goes under the name

"The Porcine Manifesto," and that amounts to nothing less than a blue-print for a new transcendental spirit.

It calls, to begin with, for something that can be achieved with surprising ease: a whole new approach to selective breeding—though *this* time with an eye toward enhancing the vastly acclaimed *intellectual* potential of the species. Now the major barrier before this at the moment is the fact that no one has yet hit upon a socially acceptable—or perhaps sufficiently dramatic—means of making that potential known to the general run of man (who wouldn't ordinarily stop to think of such things on his own).

As an *Esquire* magazine article stated with regard to dog breeding: "Beyond the pure pleasure of looking at one kind of dog, social considerations influenced breeding. If, for example, Kublai Khan carried a Pekingese, then soon the Duke of Si-kiang and eventually all the counts of Si-kiang found intrinsic virtue and great beauty in the Peke." In the case of the hog, to be sure, there's much more at stake here than his acceptance purely as a thing of beauty. Indeed, the hog is slowly becoming acknowledged—and sometimes feared—as a key component in the coming upheavals of the upcoming age.

Yet now and again there comes to the fore a man of political stature or high public standing who is unafraid to act upon the deepest, largely subconscious leanings of his immediate era. Thus if only someone in high elective authority—or perhaps even some celebrated pro football player—would stand forth and publicly embrace the hog—acknowledging that hogs and ourselves have simply been at cross-purposes for too long now, and endorsing the virtually infinite porcine potential—much of the general population would find it so much easier to follow suit, and then go on from there.

So far, so good. Can we do it? We can. The next step in the hog's ascendancy—even more obvious and only a little more difficult to effect than the first—involves something of a spiritual betrothal. Whereas marriage is what we do to define man/woman affection in this culture, man/hog affection must be defined simply by "not killing." Under this arrangement, more and more people would begin to look upon him from the light of an expanded time perspective, comprehending to a greater depth than was ever possible before that both hog and man are bone and blood and pulse and breath and persistence, each likewise fashioned on links of flesh.

As greater numbers of newly sensitized people come into an awareness of their own inner hoghood, the hog can be viewed as a basically goodhearted and occasionally exasperating mixture of skill, talent, charlatanry, tomfoolery, eccentricity, pluck, grit, spunk, aspiration and frailty. It's on terms such as these that he can henceforth be accepted.

Doubtless there will follow an "age of awkwardness" on the hog's part (a phase of spiritual and intellectual adolescence), for the hog is not faultless but is prey to the normal array of pretensions and silly self-assertions to which higher animals regularly succumb. Yet in time and with patience he will eventually emerge into the sunlight of his best realizations. We may simply need to indulge him for a while.

For his part, the hog remains a handle to hang our dreams on. Through him we can come to appreciate the full flow of lifely electricity in other living creatures. Of course, there's no getting around the fact that not everybody is necessarily going to see things this way, or even be *able* to see. As my late father was fond of saying, "Some people feel the rain and some folks just get wet." One just has to place one's faith in those who are capable of feeling.

But it is the hog, after all, who will bring about these awarenesses simply by the fact of his existence, and of his obvious persistence in the effort. In this way, he serves as an inspiration to seek out realities beyond himself, and as such he becomes a prober of the soul . . . a watchman on the city wall . . . a spiritual revolutionary and a waker of the masses who penetrates consciousness and bores to the shimmering core of life itself to shake things up and set us free.

Since status change is the essence of eventual social change, "The Porcine Manifesto" begins on the grass-roots level by calling for a *general un-ringing of snouts*, along with intensified efforts in declassifying the hog for the benefit of the layman.

As a consequence of this and other acts of liberation it should soon become luminously clear that *a custodial relationship of the present sort simply cannot continue to prevail* and is one of the fundamental aberrations of life we must, perforce, confound and put to shame. Simply the awareness of this need can be seen as the prelude to a vast reassemblage of soul, a change in the official sense of who we are and what we're doing here.

There's no immediate millennium in the offing as a consequence of this awareness; just an accumulation of simple creature contacts and linkages and inevitable irreversible understandings—maybe leading up someday to a new order of human beings. As Stanley Keleman says, "We are coming to a time when the perception of ourselves as living creatures will determine the nature of our investigations into the universe as well as our politics. We will no longer see ourselves as objects in a world of objects." We may instead come into a wholly fresh vision of creature

consciousness, a vast new sphere of post-human thought wherein we ourselves can learn to flow and function with the elemental beauty of animals upon the earth, enjoying, progressing, doing what we ought to be
doing in a creative way without any aspirations or sense of letdown.

To promote a basic familiarity and start people on the porcine path
toward a more perfect (or at least less flawed) vision of the universe, the
hog simply must be demystified and sensibly debated. For this reason,
hog encounters are encouraged in every community, with the immediate
aim of fostering little enclaves of understanding among people, including
even those who play bridge, wear neckties and swear by *Reader's Digest*.
These would be basic consciousness-raising activities and would at least
get the process started of opening up a new energy flow and tapping into

228   THE HOG BOOK

the hog current. As a by-product, these gatherings would serve to create an extended subculture of support that would lend further impetus, perhaps, to formal *porcine seminars* by noteworthy citizens, and possibly thereafter to the holding of gala *hog expositions* in major cities, conceivably consisting of multi-media displays, street theater, games, exhibits and maybe even booths in which individual pigs demonstrate unique feats of prowess.

As soon as these educational endeavors begin taking hold, the next step would be the acknowledgment and social acceptance of those countless souls who belong to what now ranks as a slightly disreputable netherworld composed of secret cabals as well as widespread half-hidden clots of avid hogfreaks and pig-addicted farmhands. With these individuals free to provide further guidance, perhaps as circuit-riding "hogurus," there might begin the formation of *hog clubs*—not similar to existing breed associations but more like socially oriented urban organizations on approximately the same level as dog or horse clubs, whose members devote themselves to the pure aesthetics of fine swine and to building an ever-better hog. These might be in the form of peer groups on college campuses, or even, for the very young, the formation of little bands that might call themselves the "pork scouts."

For the hog's private and unfettered self-development it is further proposed that the U. S. Congress establish a system of regional *Hog Sanctuaries* on government-owned land. Then too, matching funds might well be provided for the progressive replacement of the nation's more sizable hoglots, sties, pens and other such presently filth-infested ghettos with big, gleaming white *"Hogodromes"* (or smaller *"Pigatoriums"* for less-populous herds) which would essentially be comfortable rest havens, though well-stocked with intellectual stimuli and ample recreational facilities. This, of course, would be an interim move toward hogs' eventual acquisition of a turf of their own, conceivably like reservations.

In sum, the Establishment is hereby called upon to honor the emergence of a whole new swine substratum whose members—thanks to the simple gift of a full-allotted life-span—would be able at last to develop the culture and language all hogs are naturally endowed with . . . and through which they might then pass knowledge on to future generations. The land simply must be made fit for them.

Predicated upon the rapid rate of mental growth this would serve to usher in, it's entirely plausible that an ultimate upshot could be an intercultural arrangement drawing on the best of both worlds more or less akin to an *exchange student program* between young shoats and children. Moreover, if a formal educational endeavor of this nature were to prove profitable on a long-term basis to the progressive enhancement of hogs (and, too, if a multitude of other variables and hopeful suppositions were to fall successfully into place) it's not inconceivable that in the far distant future we might even consider extending them suffrage!

The aim of all this is the amassment of new awarenesses from which all good things naturally flow. It is said that to be truly loved is to be truly known by another. Thus, if we can know, love and understand the hog and what he's worth, we can begin to understand the reality behind all shadowy things. It all appears fairly feasible and attainable within the span of, say, a score of years. Many notions may fall easily into place and many may turn out to be completely impossible, but the effort itself would say something and would surely inspire future concepts that can't even be mentally formulated at the present time. As George Homans said of the nature of revelations, "To overcome the inertia of the intellect, a new statement must be an overstatement and sometimes it is more important that it be interesting than it be true."

On an immediate and purely pragmatic level, we shall need straightaway to substitute our inherited residue of malevolences and misgivings with a freshly soaring cosmic vision. This can manifest itself in various areas of human life. For example, rather than perpetuating the usage of all that foul-hearted hog terminology coined by our ancestors and grounded in ignorance, we can proceed to create an entire new lexicon of exciting new expressions and similes, such as, "slick as a pig" . . . "tender as a sow" . . . "dashing as a boar" . . . "bright as a shoat" . . . or "brave as a hog." A light-footed nubile young lass might be called "pigquick and pretty," a task well executed might be done to "porcine perfection" and a turned-on euphoric soul or someone otherwise caught up in some exuberant joy could be described as in "swinecstasy." Already the Germans frequently use the term *Schweinschen*, or "little pig," when referring affectionately to their own children. We might even go so far as to become accustomed to saying "oink" to each other as a general greeting or positive affirmation or acknowledgment.

Another such point of departure should be in the sort of ethical or emotional ideas with which we associate hogs. Rather than continuing to prolong the old greed connotations of piggy banks, we can foster the cuddly concept of stuffed toy "teddy pigs," or "rocking pigs" for children to ride on, or any number of other fresh points of reference.

Finally, and perhaps most importantly, rather than our being forced to fall back upon the hoary, overloaded legacy of yesterday's hoglore, we can (in fact, to assure our vitality as a people we *must*) devise a more up-to-date array of rites and symbols. This means far more than simply replacing old with new; it's a question of injecting fresh symbolic substance into the husk of contemporary society—a society that's daily slipping deeper into a "value vacuum" on account of its members having been encouraged to fill in meanings for their lives with *tools* and *techniques* instead of valid mythology. This is why no symbols today seem to have any compelling power. Nevertheless, humans still need symbols and organizing myths not only to relate to the external world but to work out their own identities. Obviously, then, it's necessary for symbolic expression to evolve and grow along with everything else. "The art of free society," wrote Alfred North Whitehead, "consists first in the maintenance of the symbolic code; and secondly in fearlessness of revision, to secure that the code serves those purposes which satisfy an enlightened reason. Those societies which cannot combine reverence to their symbols with freedom of revision, must ultimately decay either from anarchy, or from the slow atrophy of a life stifled by useless shadows."[2]

Hence:

A NEW HOG RITUAL FOR OUR TIME

Though neither much publicized nor widely practiced out in the open as of yet, the two-day observance of solemn commemorations, salutations, revelry, spiritual reunion and assorted high ritual matters attendant to the periodic "Festival for Re-Deification of All Hogdom and Memorial for Porkers Past" might very well turn out to be the answer for modern man's subliminal craving for fresh pig myths, as well as practical and legitimate idolatry.

Being based, as it is, upon the fact that the ordinary hog probes to our most deeply recessed guilts and fears—fears of dying, for one thing—the Festival revolves largely around the old theme of Death and Resurrection—though with distinct *new* emphasis upon the Resurrection, in league

with the accompanying belief that "it takes a hog to set us free." In this regard, "Hog" is viewed here not simply as a bastion of stolidity but as a metaphysical figure who breaks mental boundaries and challenges dated assumptions. Implicit in all of this is the acceptance of the fact that the hog, though steeped in sacredness from days of yore and invested with all manner of mythological ramifications, is, in the final act, real. We need him to keep ourselves in perspective and as a constantly renewing source of energy for our souls. Hence the Festival is an expression of the reverential awe, appreciation and mutual kinship as felt by humans, combined with the simultaneous aspect of being a joyous "coming-out party for pigs."

The occasion commences very quietly. The entire first day, indeed, is carried out in silence, for the most part, owing to the fact that preparation for the events to follow demand considerable prayer and fasting, allowing the mind to dwell upon only the most peaceful and benign sensations. The behavior of the individual participants is marked by great openness and most scrupulous courtesy not only toward one another but also toward all living creatures, every one of which—dog, cat, fly, blade of grass—is smiled at and addressed as "loving brother."

Ah, but come sunrise of the second day there erupts a mighty threshing of bells followed by the plaintive voice of a boy atop the ritual watchtower crying out: "No hogs on the horizon!" And at this the participants each begin calling *Sooey* in heartfelt tones as they rush about embracing one another. This kind of thing goes on for several hours more until another gonging of bells signals time for the Great Processional.

The prescribed route of the Processional follows the full length of a mile-long boulevard, lined with barbed wire, which leads all the way to the sacred site. The march itself opens with the onset of belaboring drums, on which note a troupe of revelers falls into formation behind the Hoguru, or Master of the Revels, who wears a brilliant headdress of bird of paradise feathers and

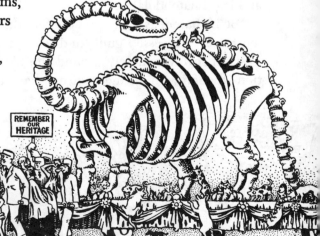

REMEMBER OUR HERITAGE

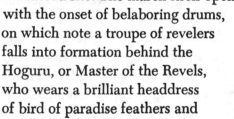

a nose spray of cassowary quills plus seven-league boots (for symbolically crossing the River Jordan). Lining up single-file on each side of him is an accompanying squad of acolytes, spear-carriers, drum-beaters, cymbal-clangers, door-openers, liveried footmen and egg-gatherers, along with a Hopi shaman and a tugboat captain. In front of them all a naked baton-twirler stands poised, wearing only a sort of skullcap with erect pig ears.

And now, upon an elegant flourish of trumpeted grunts, the Great Processional gets under way, with the Master and his retinue proceeding in stately fashion to the martial roll of drums. Not until they are well along the parade route does the main body of the Processional—which is divided into three distinct parts—set forth down the path behind them.

The first contingent is led off by two little girls in flouncy party dresses bearing a large banner which reads: THE DISPOSSESSED, FORGONE, FORGOTTEN. They are immediately followed by a solemn-faced column whose members include out-of-work clowns, arthritic trapeze artists, blacksmiths, cakewalkers, redcaps, bootblacks, coolies, steeplejacks, geeks, gandy dancers, the dozen best and still extant marksmen from World War I (in original battle dress including puttees) and a boy twirling a hula hoop while on his head is perched a big blue eagle whose outspread wings are emblazoned with the message "NRA We Do Our Part." After this comes a man labeled Captain Jinks of the Horse Marines . . . Colonel Blimp and Kaiser Bill . . . the Keystone Kops . . . the doorman of the Waldorf-Astoria . . . a flapper . . . a lollapalooza . . . a Pre-Raphaelite . . . Rollo the Rich Kid . . movie moguls chewing pink bubble gum cigars . . . a detachment of ball-turret gunners . . . a line abreast of people wearing 3-D glasses . . . and an assortment of carpet-beaters, well-witchers, snatchpurses, bear-baiters, chimney sweeps, hoop-jumpers, wing-walkers, zoot-suiters, fallen Freudians, Peeping Toms, aging surfers and asthmatic pearl divers strewing the products of their labors. Next struts a set of soiled doves from dance halls followed by numerous has-beens who never were and then by a grinning, waving drove of luminaries and personalities of stage, screen and TV

such as the Andrews Sisters, the Ink Spots, the Doublemint Twins, Fabian, Frankie Avalon and a fellow billed as "The Tap-Dancin' Devil from Dearborn, Illinois" who persists in trying to work out some routine in time with the heavy *Boom-Boom-Boombiddy-Boom* of the drums. On toward the tail end of this part of the Processional appear Joe Palooka, Amos 'n' Andy, Rex the Wonder Dog, King Farouk (working a cameo appearance), Round John Virgin, Carmen Miranda, Barney Google with the goog-goog-googeldy eyes, a pair of vintage 1950s green Martians armed with ray guns, and several stretcherloads of elderlies swilling bottles of Hadacol. Finally, bringing up the rear (and preceded by an asbestos-suited fellow holding up a sign saying "Remember Our Heritage" on one side and "Out of Sight, Out of Mind" on the other) there comes now a large float bearing a fully intact skeleton of a Brontosaurus with a pack of dogs of all sides gnawing furiously away on every available bone.

The second contingent of the Great Processional is heralded by a banner that proclaims: LIFE AS WE KNOW IT. And now, marching out amid a flurry of confetti, is a gaggle of Junior League ladies from Harlem . . . the gnomes of Zurich . . . a delegation of people from the Balkan countries, all outfitted with the native costumery they wear for movie travelogues . . . members of the Zionist-Bolshevik-Wall St. Banker Conspiracy in silk hats and spats and vests emblazoned with big $-signs, lighting long cigars with foreclosure notices . . . fly-by-night dancing instructors . . . homosexual football coaches . . . miscellaneous walking hairballs & dopesuckers, vagabonds & troubadors, flunkies & sycophants, japesters & scurvy knaves, junkies, whoremongers & degenerates . . . a platoon of young stockbrokers gnashing their teeth in

anguish and running fingers through hair that grows noticeably thinner with every few steps . . . men carrying placards marked "Pride," "Avarice," "Covetousness," "Sloth," "Failure to Honor Father and Mother," etc. . . . formidable old ladies in floppy hats with fat ankles and no necks bearing offerings of Tupperware . . . Ace, of Ace Bandages . . . Mr. Big, a rarely seen idol among underworlders who lives in Chicago . . . professional innocent bystanders . . . a trembling middle-aged man labeled "America's Favorite Basket Case" . . . scores of brash upstarts, asskissers, pop-offs, know-it-alls, whippersnappers, and White House

aides, followed closely by the Forces of Evil marching three abreast and wearing pin-stripe maroon suits with wide-brimmed hats and watch fobs. Lastly trots a band of Con Ed workmen in electric blue Edwardian outfits and ruffle-fronted shirts marching in close-order drill, chanting: "Dig we must, bum, bumbum . . ."

The third and final contingent sets forth, proudest of all, under a banner declaring: HUMILIATED BUT LIBERATED WORSHIPFUL SUPPLICANTS. This banner is strung between two poles whose ends are held in the pouches of a pair of kangaroos who hop along without any particular coordination between them. This segment of the Processional is headed by a team of fumigated hard-hats wearing garlands (*Boom-Boom-Boombiddy-Boom*), after whom come chaste maidens hand-in-hand with moustachioed beef barons from Argentina . . . then Puerto Rican youths passing out contraband postcards picturing the British and Russian pandas in the actual act of mating . . . a black-caped magician with his helper, a hulking Negro in a pink fez who wears a sandwich board reading, "Renaldo & the Punjab" . . . an army of mad poets chanting mantras . . . tiresome deep thinkers of second-rate thoughts . . . village atheists . . . and six smiling young men from Rangoon along with one bright lad who can stuff three tennis balls in his mouth at once. (All of them strutting stalwartly onward to the musical bellow-boom-gronk of tubas blaring forth the melodic strains of *The U. S. Porker March & Quickstep*, sounding somewhat similar to the *Grand Canyon Suite*.) After this appear 250 drunk men in baggy, rumpled suits stumbling along, singing, whooping, hooting obscenities and stopping every tenth of a mile to vomit and have D.T.s. Next in succession plods a phalanx of barefoot farmers wearing Brooks Brothers three-piece suits coated cuff to knee in pigdung . . . assorted Druids and samurai . . . a spontaneous chorus of old black field hands with overalls and big brown shoes chanting Hare Krishna in eyeshut ecstasy and tapping finger cym-

bals . . . distinguished businessmen and Captains of Industry leading others on bejeweled leashes . . . a pair known as "The Amphibious Two-some" . . . the Antichrist, a swarthy fellow from Peoria in a derby . . . astronauts, beauty queens and clerics . . . makers of B-grade movies walking along as if in a trance with their outstretched arms held up by helium-filled balloons on long strings . . . the Big Cheese, the yellow peril, the red menace . . . the lame, halt and aged . . . a forever unravished bride . . . a band of radical "Swine Self-determinationists" in Davy Crockett costumes accompanied by their youthful auxiliary, the Pork Scouts in fuzzy beige uniforms with curly tails . . . (*Boom-Boom-Boombiddy-Boom*).

And then here, fully resplendent in the sunshine, follow the floats, a whole line of ornate or otherwise striking tableaux towed along by tractors. The first float consists of gobs of cotton candy sculpted to look like pink clouds on which lounge a half-dozen hugely fat girls in tutus or gossamer frillies winking obscenely and striking Garbo poses. The ensuing displays feature a host of emancipated wildlife, including an orangutan in a breechclout, a toad in a top hat, a turtle princess, a panda plucking a harp, a mandrill on a horse, a chocolate moose, a lame duck, a sacred cow, mares eating oats and does eating oats, etc., a collie in a smoking jacket, a mule on roller skates, a sheep in the raiment of a wolf, and so

236

forth. Then after this there come the hogs—comes, in fact, the entire gamut of hogescence: A hog in the raiment of a sheep . . . hogs rubbed in frankincense and myrrh . . . psychedelic mushroom-eating hogs . . . hogs bearing gifts marked "Beware" . . . hogs wearing crowns of thorns . . . hogs swathed in white ascension robes, snowy as the down of angels' wings . . . rakehell hogs . . . hogs strange and wonderful . . . racehogs . . . watchhogs . . . Tennessee Walking Hogs . . . hail-fellow-well-met hogs . . . hogs of every breed, nationality and vocational pursuit—and finally, pulling up the rear, a single file of fifty porkers trudging snout in

tail like a bevy of elephants, followed at the very end by a graceful lone hog dancing on his points.

The entire body of this three-part cavalcade troops solemnly down the length of the concourse, marching into the dimness of late afternoon, to a clearing approximately the area of a football field, whereupon the participants assemble themselves in concentric semicircles around a glistening megalith and promptly commence waving wands or bull-roarers with a sense of shivering glee. For here—looming up large before them, and flanked by a pair of winged victories—stands now the FOREVER OBELISK OF HOGRITUDE, a great stone swineshrine vaguely shaped like an acorn. Out of the center of this sprouts a marble spire that soars upward to a height of perhaps 130 feet, at the very top of which sits, almost as if it were balancing there, an oblong cylindrical object roughly the size of a sofa.

The gathered faithful continue their ceremonial ovation until their noise is drowned out by louder rumblings and a mighty revving of engines as row upon yellow row of monstrous smoke-belching American-made earth-moving machinery with exposed gears and huge treads proceed, with much clanking and sputtering, to converge upon the obelisk—and promptly break down. At this signal the Master of the Revels (outfitted now with the cloth wings worn only on highly sanctified occasions) wades across the pearl-encrusted trough representing the River Jordan, which encircles the shrine, and lays a wreath at its base. Then, arms outspread, he launches into the traditional hog-directed incantation: "O root out the root of our sickness."

This is a cue for starting the *Orgy of Defilements*, a ceremony of calculated scandalization wherein the similarly chanting ritualists proceed now to roll in the trough, eat corn, run around on all fours and heap upon one another the sort of humiliations, abuse, profanity and deathful glarings that society normally levels at hogs. Gradually, though, these defilements take on the aspect of multifarious *ad hoc* acts of atonement and reconciliation, which conclude moments later with the members of

the entire entourage performing a ritual "peeling of the membranes from their eyes" as they gaze toward the obelisk and, to the slow beating of tom-toms, together cry out: "We stand with you in your selfless struggle!"

This is answered by a reciprocal "rending of the veil," whereupon the gate of the big acorn swings open and suddenly a torrential outpouring of hogs all shades of the rainbow, hogs of hope and glory with the words "me 'n you" inscribed on the sides of each, rushes forward into the open arms of the congregation. Their grand debut serves now to usher in the long-awaited *Orgy of Adulation*, signifying the revelers' mass identification and acceptance of all hogdom.

This event is marked by much merrymaking and mutual conviviality, including the prescribed dancing of the Swine Fandango, a sort of Virginia reel in which porkers and people face each other in separate lines and "do-se-do" a great deal. Thereafter, the group forms into a large spiral configuration—celebrating a seamless web of kinship and reaffirming oneness with hogness (and everything else as well)—as the participants start to sway from side to side and softly drone, "Leanin' on the Everlasting Hog."

And, ah, how the spirits flow forth. The holy moment is nigh. Ecstatic spectres gather. People weep openly in happiness. And, at last—touching off an instantaneous rash of healings, rejoicings and the laying-on of hands—an ordained hog is symbolically washed in the Blood of the Lamb. And everyone, everyone is totally energized. The ugly are beautified. The despised are redeemed. People openly disrobe, symbolic of their new sensitization and re-recognition of the elemental vibrancy that pulses within living tissue.

But now, with the sky darkening, the time has finally arrived for the formal *Resurrectional for the Porcine Race*. And now the Master of the Revels, in his role as Hoguru, drinks the potion that Circe gives to wanderers which turns men into swine and, with his face hypnotic and protean, stands squarely before the obelisk with head upraised. And now he speaks the Pork Doxology:

"*Swine O swine, so wiser than time,*
You who thrive in tempo with the rhythms of the universe
Yet who remain among us like a being long entrusted
with some sacred parlous duty here below:
We see you as the germ of future glory,
As the swollen vision of our souls' desires;
We see you as a cosmic force which flows and flickers
Like a star forever searching for a firmament;
We see you as an entity imbued with subtle secrets
That will never be revealed to any man who's not
Prepared at once to wrench his very heart
Straight from its roots, and then consume it whole
Again to be reborn."

And at this point the members of the vast assemblage, their massed faces directed toward the spire, respond in dutiful union with the authorized words:

> *"Worn and fallen at your task,*
>   *Though beaten you have conquered death!"*

—following which they commence to chant *Hoooooggggggggggahh . . . hooooogggggahh . . .* in sonorous intonation, as, from somewhere, there slowly floats the muted croon of a sultry clarinet playing "Stairway to the Stars."

Then, in his most suitable silvery caterwauling, rising above the sounds of the rest, the Master adds his final lines:

"The hope of hogs of ages past, the hope of hogs to come . . ."

And all at once, from high overhead . . . from the curious object perched atop the obelisk . . . there now drift dim flutterings and muffled reverberations like inner explosions of energy, followed next by a steady electric hum that slowly grows into a strange wavering wail of seemingly secret liquid gusts blending into something akin to a soprano-pitched swine cry. The eerie noise swells still louder as the object itself begins to quiver precariously. Now its rude exterior unravels . . . shreds . . . and splits apart like some enormous cocoon. And suddenly, breaking free from this newly-opened chrysalis, there appears the all-transforming vision of a magnificent winged thing rising: a full-bloomed hog in radiant emergence, rising . . . rising high on stupendous butterfly wings which bear the spectre swiftly upward, stealing away into the opalescent shadowlight of evening sky, ever higher: soaring on, on beyond the clouds, and then altogether past the reach of eyesight.

And in the wake of its disappearance there remains—hovering in the air, just like the Cheshire Cat's beguiling grin—the lingering, heart-cracking, soul-searing echo of a lone hog squeal . . .

X

Immanent Divinity

Like cherry blossoms
In the spring
Let us fall
Clean and radiant.
—From the concluding page of the
diary of a kamikaze pilot, 1945.

From nearly a full mile away they could see it glowing red, getting larger, clearer across the stretch of distance and clean pre-dawn darkness: BRUCE'S TRUCK TERMINAL ☆ GOOD EATS 24 HOURS.

Moon began downshifting and gently hitting the air brakes, making the dashboard cat's eyes flash with each tap. "Maaan, I got my mouth set for about twenty-seven eggs an' hash browns an' biscuits . . ."

"An' coffee," sighed Hoover, just beginning to unlimber himself.

"Coffee, yeah. An' pancakes an' pecan pie an' maybe a little waitress—to go." Moon jiggled his cigar up and down between his teeth like FDR and cut his eyes sideways. "Hee."

The direction signal on the instrument panel made Moon's face flicker green as he turned off the highway and nestled into a parking site next to the truck stop restaurant, alongside other big diesel trucks just off the road and still growling softly as their drivers sat inside the terminal shaking off the strain of nighthauling.

Almost as soon as the truck came to a full stop, the side door of the restaurant swung open and out loped a seedy, toothpick-chewing countrified face with a flattop and long sideburns. The fellow wore Levi's and low-cut boots. Altogether he carried the stained and pummeled look of a recent parolee until he climbed up into the cab of his own diesel rig and gunned the motor he had left running, keeping warm.

A cold wind slitted across the paved expanse, illuminated in this starless darkness only by the big red neon sign out front and the lights of the Phillips 66 fuel pumps. Small amber lights glowed along the edges of the half-dozen big truck trailers parked around the cafe and towering over it.

Setting the engine on idle, Moon opened the door, grabbed a handhold, stuck the toe of a worn cowboy boot into a footing and, with one smooth graceful motion, swung to the ground, ending with both feet planted firmly on the concrete.

Hoover tried to duplicate the action but wound up half falling out of the cab. "Lord, I been sittin' so long my legs turned to rubber," he said with a laugh as the two entered the cafe.

Moon pulled the frayed rawhide gloves off his hands, crumpled them up and jammed them into the rear pocket of his pants. "This is where it's at," he declared with a wink. "Best kinda place in the world." He nodded toward a sign behind the cash register that read: "Truckers Welcome. Sorry Tourist, Truck Drivers Served First."

They walked around a hard-eyed-looking woman and what appeared to be her fat daughter mopping the floor with Pine-sol and folded themselves into a booth next to a wall plastered over with yellowing photos of big tractor-trailer rigs with their drivers posing beside them. Here, too, was an amateur painting of a White Freightliner blazing down a road above the words, "Big Daddy's Alabama Bound."

In the next booth men in boots and baseball hats and visor caps with emblems of leaping swordfish and their names on their workshirts, or the names of their trucking lines, muttered quietly, occasionally breaking into little runs of laughter. At another table a plain-faced girl with hair the color of used kitty litter and a flouncy cheap pink coat draped around her shoulders sat with five loud truckers and covered her mouth whenever she was forced to laugh at some barnyard obscenity. Behind the counter an old woman leaned over a steamy sink and a black man flipped eggs in a skillet.

After the two had placed their order, Hoover suddenly announced, "We need some music," and jumped out of the booth to see what the jukebox had to offer.

Hoover felt a little out of place in this setting. He had been fairly pensive and quiet for some time now, watching these people here, all of whom seemed to know each other, or at least know about the same kinds of things. Here he was at 5 A.M., about forty miles west of Chattanooga, feeling steadily ill at ease about where they would soon be arriving.

The jukebox featured nothing but country and trucking songs: "Freightliner Fever" by Red Sovine, "Truck Driver's Prayer," "Truckin' to Glory," and "I've Got Those Drag 'Em Off the Interstate Sock It to

'Em J. P. Blues." On and on. Nothing helped: truckers guffawing in the booths; a boy going from table to table hustling Benzedrine; a pair of short-haired meaty women sitting at the counter ordering one hamburger after another; a color print just over the jukebox of a Chinese junk with an oriental mountain lake scene in the background; next to that a calendar with the message, "Make a Date to See Rock City"; and then some old cartoons and crude jokes clipped from *Parts Pups* and taped to wall. He felt very cold and alone here and finally settled on a tune without even looking at the title before drifting on over to inspect the glass display case next to the cash register. Taped to the top of the glass cabinet was a sign dotted with little grease stains:

> *"I am not a slow waitress.*
> *I am not a fast waitress.*
> *I am a half-fast waitress.*
> *Rose W."*

The case was filled with truck accouterments: flasher lights, tape players, a huge white Holy Bible, fleecy throw pillows, watchbands, blankets, pliers, cigarette lighters, $2.98 wallets, shaving cream, 25¢ Red Bird handkerchiefs, pocketknives, sunglasses, cuckoo clocks . . .

A portly waitress with an elaborate and neatly emplaced 1940s hairdo began unloading food at the booth as Hoover started back to sit down. Outside, new trucks were pulling in while other cars and trucks passed by out on the road, visible only as red flits of light fading into the distance, heading west. By this time the little cafe boomed to the sound of Townes Van Zandt on the jukebox: *"Lord, I'm gonna ramble till I get back to where I came . . . White Freightliner's gonna haul away my brain."*

"Oh man, this is the life," vowed Moon, proceeding now to plunge into a plate of steak and eggs with French fries and three slices of pie. Hoover had ordered only a tomato and lettuce sandwich along with a heavy white mug of ultra-strong coffee.

"Don't know if I tol' you," said Moon, his mouth half full, "but when I get back home I'm gettin' me a new bike, a de-*luxe* one."

"Yeah?" said Hoover absently.

"Oh yeah. I had me a Honda 750 a while back but I laid it over, went up under a lady's car. I 'bout like to got killed! Didn't I tell you about that?"

"No."

"Hell, I don't even have a kneecap. See?" He pulled up his left pantsleg to show off a set of deep scars. "I 'bout like to got . . ." Moon fell silent and froze a full second as something seemed to shift very slightly beneath the superstructure of his face. Then he slowly lowered his bell-bottom trouser leg over his boot. "You know," he said after a pause, his voice a little softer. "I understand how you feel. I don't like this any better'n you do. You know this ain't my kinda thing at all."

Hoover pursed his lips, puffed out his cheeks a little and nodded.

Then Moon reached over, clutched his friend's knee and, looking straight into Hoover's eyes, said without the slightest twinge or undertone of his usual devilment or whimsey, "Let's not leave 'em out there all alone any longer."

At this, the two men rose, left a tip along with all the uneaten food and walked swiftly to the cash register. Behind the counter a boy was leaning in a chair against the wall poring over a men's magazine. Above his head hung more faded, curling truck photos plus a picture of Jesus praying in the Garden of Gethsemane plastered to the wall alongside an assortment of fan belts and radiator hoses.

The boy looked up with a start, then sprang to his feet and rang up the check on the cash register. On handing back Moon's change he darted a quick glance across the room. "If'n you're needin' to stay up much longer," he said, "I got a good supply of Ben Whites. I'm sellin' 'em by the number. Give you a good price."

Moon stuck the change in his wallet shaking his head. "Naw," he said quietly, turning to the door, "got all I need."

The overcast sky by now had lightened into the first bare cold grayness of dawn, and greater numbers of trucks continued to lumber in off the road: tractor-trailers, semis, flatracks, doubles, cattle rigs; Kenworth, Peterbilt, Ford, White Freightliner, Autocar, GMC, Mack—all with their different payloads: produce, chickens, livestock, freight, furniture, oil, hay, explosives.

With Moon's and Hoover's reemergence into the air the cargo of pork on the hoof seemed now to reactivate itself with murmured grunts and restive shiftings.

Back in the cab, Hoover listened to them a moment, then rolled a fresh cigarette, put it in his mouth unthinkingly and gazed out the window in silence. Meanwhile, Moon pulled out some paperwork and hastily tried to put his driving logs in order—figuring the time and making up hours in order to gain a balance—until he noticed Hoover sitting there. He reached into his pocket to fish out his lighter, tapped Hoover on the shoulder and lit his cigarette before unwrapping another cigar for himself. This done, he released the brake and commenced now to back carefully out from between the other trucks and maneuver back onto the highway with the deep diesel rumble of his rig changing pitch as it picked up momentum.

The road careened out across the landscape long, smooth and dead flat for as far as they could see, with, just now, the first few drops of a rain beginning to dot the asphalt.

After a few more miles, Hoover broke the silence. "Listen to 'em," he said, exhaling a lungful of smoke. "They're walkin' about."

Back in the truckbed, the hogs were swaying softly together, crammed and jammed somehow even tighter than before, grunting and murmuring nervously among themselves as if communicating now on some entirely new frequency. Their voices sounded like atonal oboes. Through the slat sides of the truckbed the sights along the roadside, growing lighter gray with the morning, seemed to whir away from them faster and faster than ever.

And soon they were into the farthest-reaching outskirts of Nashville, and before too much longer they felt the truck swing off the main highway, then move along a series of smaller, rougher roads that made it harder to maintain footing, and finally turn down a long, curving roadway, traveling more slowly than before, possibly because of the rain.

At last the big diesel pulled into sight of a large stark structure standing out against the grim light of an early morning rainstorm, and then, bouncing over a speed break, it moved past the gate of the fence that surrounded that structure, ultimately coming very quietly to a stop alongside a ledge at the rear of the building. The sound of the engine suddenly died away, after which Moon and next Hoover left the cab and disappeared inside for a time.

In a short while they came walking back out on the ledge with another man wearing a helmet who slid back a huge metal panel exposing the insides of this building. The side doors of the truck now were opened and a rampway with rails was laid between the truckbed and the ledge, leading into a concrete-surfaced pen just inside. Other hogs were already gathered in other holding pens and still other hogs, in groups of threes and fours, stood waiting in pickup trucks outside, unprotected from the rain.

Hoover went back into the building at this point while Moon and the man in the helmet began to herd the cargo down the rampway, counting each of them off:

"Thirty-five, thirty-six, thirty-seven, thirty-eight, thirty-nine . . ."

The hogs trundled out one by one in perfect complaisance but increasingly apprehensive, their ears all perked, alert, attuned. Presently,

Hoover reappeared with some papers and in a few more minutes all eighty-seven of the porkers were accounted for and enclosed with a clang behind tubular steel rails. Moon pulled the truck panel closed as Hoover shook hands with the man, turned and walked straight to the çab and boarded. A certain opaque blankness had closed over his face.

Now Moon shook hands and climbed back into his driver's seat. He started the engine, revved it a bit and squinted upward through the windshield at the rain now pouring in torrents. "H'it's a real frog strangler," he said, as he began slowly pulling away from the side of the building, catching sight only a few times through his big side mirrors of the creatures he had hauled through the night, many of whom were looking directly at him through the rails of their new pen.

And in moments the truck was approaching the gate, and then past it, shifting gears, moving smoothly off down the long, swooping roadway, growing smaller and softer and dimmer in the rain. And now the helmeted man, carrying a clip board under his arm, pulled shut the big panel with a metallic boom.

*　　*　　*

". . . and the Lord shall lay bare their secret parts."

ISAIAH 3:17

Here is a calendar with a photograph of Stan Musial signing autographs for some little boys beneath a sign that says, "ENJOY OAK FARMS MEATS—45 Quality Meat Products." Below the picture is the calendar for the month of December. It is a little more than a week before Christmas and the Crowe Bros. Meat Packing Plant here in Nashville is operating at full capacity, peak efficiency.

A low whooshy sound reverberates throughout the long halls on the second floor of the building—that, and the less distinct grumble and moan of massive machinery coming from somewhere in the lower bowels. Here, too, floating in the air is the sporadic intrusion of steamy jackhammer noises and the interspersed whine of power saws and the half-drowned-out sound of men.

This is the cutting room at Crowe Bros. where about twenty workers have been laboring away at their individual tasks since seven this morning. The men, blacks and whites, wear blue coveralls with yellow rubber ankle-length aprons and white visor hats or plastic helmets. Each one wears a chain around his waist from which dangle whetting instruments and an assortment of large knives. Two men operate power saws.

The hogs enter the automated butchering room on an overhead conveyor chain, suspended on coathanger-like hooks from the Achilles' tendons of their hind legs. They arrive as emptied carcasses already split open down the chest and headless. Their dead, smooth flesh is long since hard from hanging overnight in the plant's huge 25-below-zero blast freezer to which they were conveyed yesterday one by one after being killed, de-haired and divested of their vital organs.

On entering the room they pass first to a large blade that splits them into two mirror halves, then each goes onto a wide moving conveyor table on which men, facing each other, take care of their own special cuts. Some pieces of hogmeat are placed on smaller conveyor lines to be carved up by the craftsmen in charge of those particular portions. The hams come off first and go onto their own line; the shoulders are removed and go onto theirs. The loin is trimmed adroitly by a worker with a long drawknife. Off come the ribs. Off comes the bacon. Strips and wads of fat are flung into vats on rollers. The men on the line tease a young black fellow hauling a cartload full of heavy hams. "Woooo-eeee. Here come Santa Claus." A female worker off to the side, where she operates a grinding machine, playfully hits a man with a piece of hogmeat. They all laugh.

Flip . . . flap . . . plap. Bits of hog are rhythmically tossed from one table to another or onto a smaller conveyor or into a cart. *Zing . . . zing . . . zaaaaap.* The man on the vertical saw cuts off a foot, a section of ankle and makes one long slice across the top of the foreleg.

The room is brightly lit but antiseptically cool and covered with pastel green tile. The tile floor is slippery with watery pools of diluted blood. At various spots around the room stand large metal carts of tails, of feet, of fat. All the vital parts come asunder here in this site of the swine's final disassemblage: sawed, sliced and dissected into separate parts ready for packaging. Now and again the men half sing or chant or laugh among themselves in the manner of laborers on road gangs or slaves in old Southern cottonfields. And yet there is the sense of being in very skilled company with these men, all making quite deliberate surgical-style incisions on their particular pieces of flesh, pausing every few minutes to sharpen their blades on the whetting instruments dangling from their waists. They operate with great coolth and precision and confidence, oftentimes interspersed with distinct flarings of individual style like cooks in Japanese steak houses. *Zing . . . zing . . . zaaaaap.*

"Now you take these tails," says Wilbur Gray, the white-helmeted plant supervisor who has stopped in to check on this morning's operations. "They may not sound like very palatable things but they have a real good flavor." He picks up a handful of foot-long tails lying in a cart. "They're boxed and sold. People like 'em."

Fat, bones and other miscellaneous parts, he says, are sent downstairs and boiled down. The grease is used for soap while the rest goes into high-protein livestock feed. Parts inedible by humans are used in dog and cat food. Blood is collected and dried and sold to glue factories or plywood plants. Manure is used for tankage. "Some companies buy the hair and use it for upholstery stuffing. We also send some to Formosa, where it is graded and made into paintbrushes and toothbrushes." Mr. Gray smiles and pauses for effect. "The only part we *don't* use," he proclaims in the spirit of a man who has used the line uncountable times before, "is the squeal."

Flip . . . flap . . . plap. And now a bell above the door rings in two five-second bursts, and some of the men cheer and the whining power saws shut down. It is lunchtime.

The cafeteria here at Crowe has an uncomfortable atmosphere of gloomy drabness—and almost total colorlessness except for the green tile floors which are slippery with the animal blood and bodily fluids tracked in on the men's soles. The food is chosen and prepared with minimal adequacy. No one would want to shirk work to spend much time here—except for the fact that they *do* have a good selection of vegetables. The men and few women sit at long plain tables and don't really say much—don't say anything at all about their jobs. Pinned to a bulletin board are a few printed notices having to do with company policies on safety procedures and how not to get caught up in the machinery. There are three "car for sale" signs, a wedding announcement and a printed piece about the size of a bumper sticker that says "Happy Holidays!"

As they finish eating, most of the workers begin drifting off in twos or threes, not in any apparent hurry to be anyplace special. Some of them are still sitting around, but it is quieter now and I become conscious of the gruff purr of machines coming from somewhere inside.

I leave the cafeteria and wander alone down some clammy tile halls to a short flight of descending stairs, then along another large corridor and then into a 20-foot-high room with big freezer doors along one wall. And here finally—through a rectangular tiled entranceway—are the machines, already at work.

This is a vast, high-ceilinged room, perhaps 150 feet long and 70 feet wide, consumed now in the loud rushing sound of motors and mechanisms, punctuated with irregular clanks and slaps and unseen collidings of metal. An occasional animal scream will rise above the general wailing passage of the disassembly line. There are no voices. Off to the side, a man yells a couple of words to another worker some distance away. Otherwise there's just the great tidal hum of all this machinery—or actually this one continuous machine, which fills the room.

Over in the left section of the room, near the corner and just through the entranceway, the hogs are rising up from some lower place on a conveyor with moving sides made of slats of wood. The sides look like tank treads and they are slanted inward along the bottom edge so as to wedge the pigs in and move them along.

They are rising one by one to the top of this conveyor—some of them struggling, some just looking quizzically about—at which point a black man in a helmet, yellow apron and rubber gloves touches their foreheads with a two-foot-long black device connected to a heavy electrical cord. He lays this thing upon them in a careful, vaguely ministerial manner, whereupon they instantly stiffen. Heads jerk back, eyes clamp tightly closed and forelegs thrust straight out over the bottom edge of the conveyor.

There are eight men visible from where I am standing just inside the entranceway. Two or three of them I remember to have worked in the butchering room before lunch. After lunch the chief energies of the plant are directed to this place, the killing floor. The temperature here is held to a chilly 50 degrees.

The hogs ride up on the conveyor to where they are electrically stunned and then slide on rollers down to a moving platform about ten feet along and six feet wide. There they are stabbed in the throat by one man while another on the opposite side of the platform attaches to the hog's back foot a chain dangling from a larger overhead link-belt conveyor line. The hog, in his final second or two of life, is then lifted up into the air, still dribbling blood from his nose and throat, and ferried through a long tank of boiling water. The entire overhead conveyor line winds around like a flattened-out backward "S," with the killing platform and the tank being along the top line. On emerging from the tank, the hog goes around the first curve up to a point overhead where a worker detaches it from the chain. It is then placed inside an enclosure containing two big revolving brushes which spin the body over and over and beat off all the loosened hair.

As soon as the hog comes out of the de-hairer, a pair of workers attaches its hind trotters by their Achilles' tendons to a double-ended hook shaped like a coathanger which hangs from the conveyor line. Suspended from this, the animal is hauled around the next curve to where it passes through two columns of gas-blue flames that singe off the remaining hair and harden the flesh. After this, the carcass passes between banks of spraying water, like a minute car wash, and then onward through some smaller rotating brushes, finally slowing down at a point where a waiting man cuts out the pads between the toes and any rings that might be in the nose. Next to him is a man who shaves off whatever vagrant patches of bristles remain. Following that man is another worker whose sole job is cutting off the closed eyelids, revealing the eyeballs staring out dark and deadly cold. By this stage the hairless revealed flesh is pale and pinkish white and glassy smooth like soft porcelain. They all now have the appearance of large porcelain images of hogs with no connection to life whatsoever except for the few drops of blood occasionally still trickling from the silver dollar-sized holes in their throats.

After the eyelids are sliced away the hog next swings along to a man who cuts the throat from ear to ear and "drops the head," letting it dangle forward on a strip of flesh. At the adjoining work station the head is removed and placed on another conveyor line while the rest of the body is shunted along to a man who makes a long slice down the breastbone. The guts are pulled out, the innards unraveled and placed in large "visceral pans" for other laborers to handle. A meat inspector for the State of Tennessee cuts out certain glands and checks them for tuberculosis and an assortment of other diseases.

It is here that the head and body go their own separate ways, each passing along on its conveyor into the next big room.

"Now you see," points out Wilbur Gray, who has just made his supervisory appearance on the killing floor, "here is where the hog is stamped with an inspection stamp. The kidneys are taken out, the insides scraped out. Those two men on the table over there are taking the stomach and the liver and the heart off, opening the stomach and washing it, preparing it to be scalded. The heart comes down to this work position," he nods to a table a little closer, "and is slashed. This next man spots the liver. If it has any white spots on it we cut them off. The roundworms cause those white spots. If it's got more than five we have to go to inedible with it. Abscessed livers are condemned. The man takes the spleen and the various offal and puts it on a shelf to be taken to the cooler."

Meanwhile, the head—which is already fairly cooled after being dead about sixteen and a half minutes now—twines along the route to its first stop where the ears are cut off and tossed into a blood-filled vat. Then a gloved man chisels the cheek meat loose and trims the head meat before passing on the task to another worker who stands at a table on which rests an upright metal wheel. The wheel is perfectly smooth around its rim except for a single pair of small curved metal hooks jutting out. This man holds the hog's head up to the rim of the wheel with its nose pointing to the ceiling and then steps on the motor switch that makes the wheel revolve around to the point where the two hooks go into the nostrils and rip the entire snout off. After this, the final worker in charge of the head's disassembly takes the meatless, earless, snoutless skull with its two bare eyeballs gaping blankly out of their lidless sockets and puts it on a support like a footrest at a shoeshine stand whereupon a wedge descends and splits the whole thing in half. The brains come out and the tongue and lastly the pituitary glands, which are perused by another government inspector. The eyes are tossed into a special bin for tankage; all the other scattered shards of hogflesh are put in various containers to be chilled and later processed in the sausage kitchen.

"YAWP—" the cut-short sound of a hog hit by the electricity echoes shrilly from the other room. Now another one.

"We bleed the hogs for three and a half minutes," Mr. Gray recites. "Then they're in the tub for eight minutes. It takes two minutes to go through the de-hairer and onto the chain, then about twelve minutes on the chain and they're going into the cooler. The cooler is upstairs. They're chilled overnight and we begin cutting at seven in the morning. We're gonna kill seven hundred and sixty butchers and eighty-six sows this afternoon."

We walk back out onto the killing floor and Mr. Gray nods toward the shocker in the hands of the worker who first greets the living hogs coming up on the conveyor. "That shocker, see, knocks the hog out so he doesn't feel the blade going in. The heart has to pump the blood out, so we don't dare kill them immediately. The sticker has to cut three main vessels—both main arteries and the jugular vein. If he mis-sticks one, that hog will inhale tank water and will drown. He'll asphyxiate. The veterinary will catch it out here on the rail and that hog will be tanked. That doesn't happen very often but if he does a poor job of sticking, we're in trouble. The tank water goes into the bloodstream and we lose that hog. You can tell it in the liver and kidneys after you take the guts out."

A hog is waiting at the top of the conveyor, but the machinery has stopped for some reason. There's a great restless quiet across the killing floor, an eerie tranquillity. In the corner of the room, right next to the opening for the conveyor, is a door and about ten concrete steps leading down. I have come down here behind the scenes where the live hogs are shuffling about, waiting for curtain time. Here is a line of tubular steel holding pens as long as the building wherein languish the hogs, who've been graded according to size. In addition, there are a few smaller pens back up at this end where the conveyor begins.

The conveyor is fifteen hogs long: That is, there are fifteen hogs waiting in it right now while the machinery has gone so strangely silent. No one seems to know when it will start up. There are twenty more hogs in a railed funnel chute waiting to move onto the conveyor. They have had to be driven into this final pen by a teen-aged boy with an electric cattle prod.

The conveyor goes up about ten feet and then the slanted wood-treaded sides close in on the hog until he gets to the top. On the conveyor are, as I said, fifteen hogs standing here, waiting, some relieving themselves, some sitting on their rumps, propped up by their forelegs, some majestically lethargic and some with an almost tactile sense of dread and tragic intuition rising off their backs as they sniff and catch the odors. The hogs talk among themselves—those on the conveyor and the others in the chute behind them—talk mostly in low grunts drawn out at various lengths, grufflings and quick squeals, an occasional bark and sometimes sudden screams of dawning fear. The place reeks the faint aroma of hallucination.

Off in the corner back wall is a railed pen with the sign, "PEN 26 Tennessee Dept. of Agriculture SUSPECTS," in which a cluster of hogs is huddling darkly in a spirit of tension and menace at the far edge of this building's cornermost enclosure, all of them breathing in deep sighs. They each wear large tags clamped to their ears that say: "SUSPECT (and a number) *This tag shall be detached only by an employee of the Tenn. Meat Inspection Station.*" A suspect is a hog that the inspector decided was not normal. It may just be lame. "We tattoo them and kill them together, kill them last," Mr. Gray explained earlier. "The veterinary has the opportunity to be right on the line for real close inspection. None are left alive. We buy them subject to inspection. We buy only top-quality hogs, but occasionally you get one like that."

Right next to the suspect pen, a little closer to the conveyor, is another animal with a suspect tag on its ear, but this is a cow, a very obviously sick or internally damaged cow, absolutely jet black and mooing in doleful long-drawn-out groans. They kill cows here, too, though not the same way nor with the same machinery that they kill pigs. This cow, so a boy back here tells me, was brought in yesterday and has been under observation. She appears to be on the way to dying but no one yet knows of what. I heard her moaning earlier out front whenever there was a lull in the hectic nervous scherzo of the machinery.

Next to the lone cow's pen is, again, the chute and conveyor. There are some churning metal noises outside. I can see a little bit of the killing floor through the opening in the wall for the conveyor. The place is being sprayed out with steam hoses. There is the hollow echo of some voices, and up next to the conveyor the electrical man is smiling at someone I can't see and smoking a cigarette. Straight across the moving table, the "bleeding platform" on which the hogs are cut, the sticker—who wears thick gloves and an ankle-length yellow apron spattered and smeared in red—is whetting all the knives on his belt.

"They've had a breakdown," says the boy who drives the hogs into the chute. "Start up any minute."

Looking along the conveyor, here's a neatly funneled-in line of hogs, all panting expectantly with palpable tensions, all their ears perked up now, ever since they heard the first few fresh clankings and sputterings out front. Up at the top they are jammed in more tightly nose-to-tail, though with some room to raise their heads. In the chute and lower part of the conveyor, before the wooden side conveyors close in, they can sit or even turn a little ways around to speak to whoever's behind them. Here's a curly red-haired one, must be a Duroc, flexing his nose a little nervously but otherwise just standing as serene and composed as possible with his ears flapped over and his chin held high as if trying to give the appearance of some inner calm in the face of what he obviously already senses. A light brown one with a white band across his shoulders and front feet has his snout almost straight up, looking about from side to side and muttering *"ungg . . . ungg . . . ungg"* at five-second intervals. A Hampshire gazes off to one side for a few moments, then off to the other, and sighs a lot. Others mumble things. The cow moans. As I start back up the stairs near the cow's pen I see she has gone from standing upright to being now half crouched on her knees, continually drawing her next to last breath of life. I can just barely see her eyes, she's so black.

Here it goes. Here's a grind and a clank and now the whirring rushing hum of big motors. Here come the hogs up here. Here's one, and he reaches the top of this conveyor and he looks quickly to one side and then the other and his nose pulsates and he's beginning to sense something overwhelming, and now the electrical man touches his head with the stunner. Sparks jump up from his forehead. He stiffens and slides down the rollers to the platform while another hog, a black one, appears at the top, and sparks jump on his head, too, and a little wisp of smoke, and he falls rigid onto the platform where his mouth begins opening and closing, and he starts quivering. The man puts a chain noose around his leg and the sticker plunges the blade into his throat and his eyes start to roll back and his mouth gapes open and open and open and his legs are galloping away as he lies there on his side gushing blood in a four-foot

Hog autumn: With awareness buried deep in his face, this porker now reaches the end of the conveyor that has brought him up here for his life's first glimpse of the packing house killing floor.

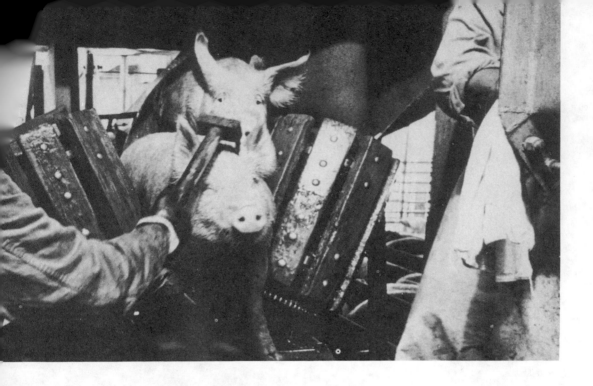

On being electrically stunned, all consciousness ceases, as the heart beats on, pumping out its last drops of irreplaceable life.

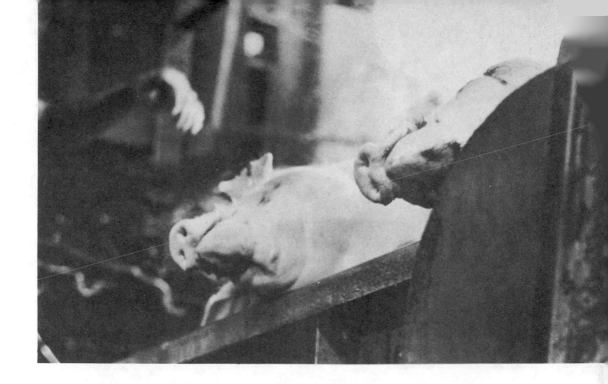

Once de-haired, they are hooked onto the disassembly line, to be parceled into the wads of meat we recognize in stores as "hog."

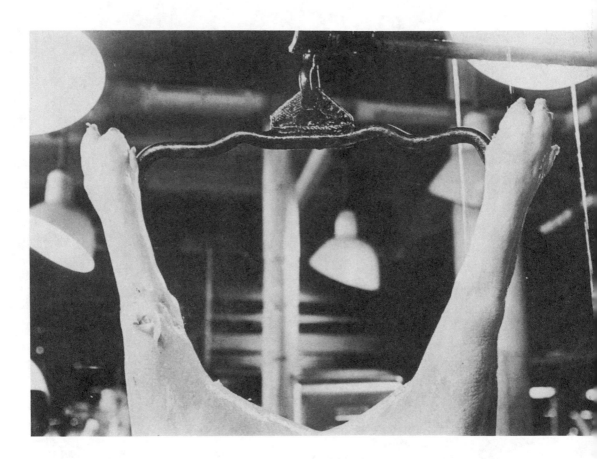

Of the fragile debris hogs leave, nothing is discarded. An onlooker with a possible flaw will be done in after all the rest.

spurt. Now he is hoisted into the air by the chain attached at its top to the overhead conveyor, which looks like a giant bicycle chain and whose noisy passage blends into the shimmering cascade of sound throughout the building.

Here comes a brown Poland China, his squeal cut off by the electricity; and here's another one. Their heads all jerk back exactly the same and their eyes slam closed. There are sparks and smoke and the odor of singed tissue. Here comes the red-haired one, big red-haired Duroc. He's looking around here, his legs all wedged in and hanging floppily over the lower edge of the conveyor. He looks hard over to his left at the man who's going to attach the chain to his leg and he looks at the sticker and he looks at the electrical man who is operating the conveyor switch with one hand and holding the stunner in the other, and his eyes open up as wide as they can ever get. And now he's electrocuted. Sparks jump. Mouth opens. He's stiff now, trembling, quivering. Now he's cut. Now he bleeds like a liquid-filled balloon deflating through a leak. His eyelids are fluttering. The chain is attached to his foot. Now his back legs begin to kick. Now he begins to feel pulled up. His legs jerk and gallop more and more frantically. And now he's beginning to be lifted into the air, this red one; he's lifting, lifted, ferried along by the chain off the bleeding platform just as the next most-recently-stabbed hog begins kicking him. Now the red is in the air, and he's still. No, he kicks a little. Now he's still. Now he's going over the edge of the boiling water tank.

Here are two hogs coming up the conveyor who are almost on top of one another: a brown one, he's gotten it now, and a black one, and now he's touched; and now a third, a Yorkshire who's twisted over lopsidedly in the conveyor. He's looking suspicious; his eyes are squinting. He hasn't been touched yet. Now he is. He's stunned. Maybe now he won't feel his throat cut. They make no noise after they're touched. Sometimes they kick and jump up in the conveyor, seemingly trying to scamper over the tops of the ones in front as if in anticipation of something pleasant on the other side, and then they're touched and their consciousness ceases. From here on out it's just the automatic thrashings any animal goes through to stay functionally alive. Perhaps the inevitability of death is indeed so built into some hogs that the notion of termination strikes no greater dread than it does; yet each of these creatures still fights with wild mindless persistence to extend its duration and cling onto its own lifely fluids.

The endless quest for life ends here: swine vibrations steadily terminating and turning cold amid the turmoil and heated din of all this machinery. This is the seventeenth of December. By day after tomorrow some of these hogs being killed will be salable meat products, including big hams all ready for Christmas.

There is the constant clattering grind and squeak of motor-driven movements and an intermittent chorus of high, wild raggedy shrieks rising over what has become a steady wavering wail coming up from the lower regions back where the conveyor starts. Maybe some of them squeal when they feel the electricity passing through the body of the hog in front of them. Sometimes they crawl up over one another. Their terminal screams are totally bloodcurdling, each one a screech tearing straight upward in the throat.

A white worker walks over to the sticker and fishes from his back pocket a pouch of Red Man. He takes a wad of tobacco from the sticker, who is black, and now he's chewing it and watching hogs bleed on the killing platform. The sticker makes his incision deep into the hog's throat with the confident swift precision of a Zen master. He is not violent but wears an expression of cold, tragic scorn. His forearms are drenched with blood.

Hogs on top of hogs. They come up the conveyor and they are stunned in quick order, first one, then the other. Here are two hogs that look like twins. The sticker sees he almost missed one and he leans over and rips his throat. Here's a huge Hamp, five-hundred-pounder. Blood gushes out as from an opened faucet, covering the sticker's feet, covering the platform. Beautiful scarlet red. The wall is splattered with it and it brightens the floor and hangs in droplets and clots and streaks like stalactites from the stationary metal parts of the bleeding platform. The light bulbs along the lower level of the wall on the other side of the platform give everything around it a fuzzed red glow. I'm surprised this torrent of blood—clotted, jellied, coagulated in red glowing smears and streams—doesn't foul up some of the machinery, or these motors. It is a richly beautiful crimson hue, almost incandescent at times; and when several hogs are bleeding away at once the blood splashes against the back of each thrashing porker and ricochets in little sprays up into the air, and the lights behind seem to turn it all into a floating blood fog—a fog that momentarily shrouds the cumulative numbing horror below. I wonder how much blood is contained in a hog.

Dimmed in this red haze, the whole procedure takes on a ritual, almost sacramental character: an authorized brutal ballet, an adagio of death wherein each performer knows precisely his part, and how to play it over and over again. (Here is a sudden scream from some dim beast with opened veins. Howls of pain and madness echo.) These are the men to whom we officially delegate our killing; and they, in turn, go about the performance of that task as if it were a sanctioned and vital rite, a liturgi-

260

cal Promenade with Knife: processional slaughter enacted with the clean brutality of a natural phenomenon. The men abstract themselves even more deeply into this ritual of respectable carnage by becoming part of the surrounding machinery and moving to its rhythms: shock, stab, chain . . . shock, stab, chain; a letting of blood; a climax of cruelty; the first step in the formalized mechanical relegating of vibrant creatures into disconnected, disassembled wads of meat. On our behalf these men take upon themselves that special etherizing of the heart which makes it possible for creatures to endure the rote extinction of other creatures. (There is an animal scream coming from somewhere back in the machinery.)

The electrical man lays the stunner on the hog's forehead with indifference, for the most part, but sometimes with a sort of solemn air like a high priest touching a brow in blessing. Sometimes, if they begin to kick and thrash too much after they've been stuck, the electrical man will touch them again, which stiffens them out and closes their eyes again and allows the blood to spray out in peace.

One hog has just wrenched himself out of the conveyor while the electrical man's head was turned and suddenly finds himself standing on all fours here on the bleeding platform as it moves along, poised to run but not understanding what's going on or how to get off this thing; and now the electrical man reaches over and touches him and he explodes into paralysis and falls over, and now he's stabbed and chained and his eyelids flutter and his eyes roll farther and farther back in the head.

Now here's a quick succession of four hogs, all cut, mouths opening and closing, jerking their legs together like a chorus line with throbbing spasms. The men move aside almost sympathetically and let the hogs kick and shake the blood out of their heads. Some hogs get "sick to their stomach" when they are hit by the juice and vomit on the bleeding platform.

When they reach the top of the conveyor, many of the hogs light up with the abrupt realization of what before has always been dimly comprehended, and their eyes widen and sometimes their mouths open without sound, but usually not.

Eskimos love and respect the "dear dead beast" that dies to feed them. Here, the beasts are killed off with a simple glib efficiency like components in a system, component parts whose lives society claims a right to. And they are killed in concealment so as not to jar the sensibilities of those for whose benefit all this is undertaken—people long since programmed and conditioned to focus on the product and ignore the process.

Yet these creatures still belong more to the earth than to machines.

They do not come into the world, they come out of it, hewn from the same organic stew that stirs all life. In tenderness they come up out of the ground, out of the plasma of the earth, and remain a part of it, their eyes never drained of hope. But early on they find that they are victims of an unappeasable power, a blocked-off, calibrated, mechanical way of thinking that ultimately culminates with their being moved along onto the actual mechanisms of places like this, caught in these inescapable cogs.

Here's a white barrow trying to jump out of the conveyor. He's somehow gotten almost over on his side and is now wildly jerking. Now he's at the top and he sees the source of his darkest premonitions and is struck—in a hallucinatory suspension of moment—by the sinister inevitability of it: the dawn of death . . . the purest fear upon the earth.

At that instant before the great strange chill, what is it that possibly transpires in that zone between the world and the mind? Does he flash back to the farrowing barn sucking sowmilk, or to a favorite field, or a kind of food? Did he ever once get to glimpse a glory in the mind to return to? A moment of ecstasy? Winging dreams? He is all alone now with his little life of some six months and squeals because it leaves him, a scream the sound of tearing metal. All at once he grows hysterical in this first-and-only face-to-face encounter with a kind of terror that is both familiar and never-before-felt. Fear permeates the meat of a slaughtered animal. And now there's an explosion of purest white voltage and . . . doomcrack. He opens his mouth gasping; he chokes and shivers, lyrical with pain; eyes rolling slowly back in his head, fading into lifelessness; kicking again; opening his mouth unable to speak; eyelids fluttering, closing; blood leaving, bubbling out; now turning cold from nose to tail.

Stone dead, lapsed life, signifying nothing; and he will leave nothing behind—except, of course, our sense of wonder. No notice of his passing, no return to dust, his body not interred back into the pig-loving earth. Soon there will be nothing whatsoever left that could possibly be laid to rest, and there is no hog Valhalla.

Dogs and cats and pet chimpanzees and horses and parakeets are buried and nobly commemorated at Forest Lawn and other cemeteries, but with hogs it's total liquidation. They go to their final sequestration in the form of whatever fractionated parts can be shrink-wrapped, canned, bagged, pickled and eventually put inside us. They have no graves from which to speak to us—other than our own bodies. We incorporate them completely, and they are us. Everyone is deeply implicated.

The cow mourns in agony from behind the scenes. Hogs bellow in protest on their ride up the conveyor. Everything struggles to survive, no matter what the odds, no matter what little their lives may consist of or how bleak or bereft of promise. When the time is clearly at hand the lust for survival erupts out of some unspeakable apprehension or perception forged deep in our furnace of blood.

Here's one hog who's waiting in the conveyor, which is momentarily halted. His nose is poking out over the edge. He's almost the same color as the wooden sides of the conveyor. He looks to one side and then the other. He looks over with stricken eyes and sees these other hogs kicking and catches the fresh deathodor and sees the men, and now unleashes a sharp, rending, heart-freezing cry from that place inside that no one who has not been prepared to visit can ever know. Now sparks crackle from his head and he slides down to the bloody metal platform, twitching and jerking like fresh-skinned froglegs with salt on them.

Surely the earth's saving grace must be that passions dim and old pains cool and hurtful visions fade into the recesses of the mind. But they are always still there. The question for later becomes how to look back on these things. For now it's how do you look upon it at all?

Everything is One, it's true, but that really doesn't make much difference. Not at this moment when you are just one single entity that's drawing to a close against all energies of will or passion or reason, awash in the monstrous just-so-ness of the world and fighting frantically against it to the last flicker of pulse. Survival's all there is.

I go back through the door next to the conveyor and down to the holding pens. I haven't heard the cow's low groans for a little while now, and as I descend the stairs I see her coal-black body lying flat on its side breathing deeply, heaving painfully, all the while emitting thin moos under her breath, and persisting in these calls and clinging onto her elemental vibrancy as long as there is any spark of being left whatever. Now she stops. I feel her inner stirrings cease. Silence.

Meanwhile, the air thumps and percolates with the terminal shrieks of hog after hog after hog. Lights going out. Life oozing out—lives that will nevermore appear but that still battle to the last against that certainty. And suddenly—back here where the machinery begins—I remember, and I become utterly, coldly terrified for these hogs and wish I could only somehow pray for them, for the cow, for my father—and for myself.

I go back up the steps to the killing floor with one short glance over my shoulder at the hogs in the chute. It's incredible that they should exist at all, or have ever existed—all so finally fragile and so easily called away, shortly to disappear from the face of the earth altogether. And up here on the floor are these men, each operating amid all this fiercely precisioned and demented machinery, though equally vulnerable, fearfully transient, totally alone and capable of comprehending it all so directly. Those who are exposed so relentlessly to things that are ending should surely know. The perception is the realization. And again I remember, and I feel monumentally liberated and reconciled in knowing that the outward reach is endless. I am released and am free to care to the fullest intensity of my being about living creatures, free to love, to revel in the joyful reality of animal warmth and the heartbeat regularity of the sun and the semen odor of fresh-cut grass and trees and water and all things that shimmer with elemental vibrancy—all of it absolutely astounding, freshly radiant and dreamlike, completely unrepeatable and worthy of whatever small reassurances can be shared. I am, in fact, alive, and openly acknowledge in myself and appreciate in others the awesomely terrible frailty of that condition, and marvel at it.

I am standing below the de-hairing machine into which the dead hogs go for two minutes after having their bristles scalded in the tank. And here is a large hog lying horizontally between big, coarse, rotating brushes spinning him rapidly over and over and giving his skin the sheen of porcelain. He is spinning almost over my head. Water and a few droplets of blood sling off the body and fall on me. Now he slides out and is suspended upside down by his back tendons, ready to move on. Ah, look at you, hog. Uncompleted symphony. Terminated aspiration. Sense of thwarted greatness. Numb and meaty visage. You were beautiful!

Notes and Bibliography

Notes

CHAPTER I

1. F. C. Sillar and R. M. Meyer, *The Symbolic Pig* (London: Oliver and Boyd, 1961), p. 83.
2. Prince Maxmillian of Wied, *Travels in the Interior of North America*, trans. by H. Evans Lloyd (London: Ackerman and Co., 1843), p. 62.
3. W. Scott Elliot, *The Story of Atlantis and the Lost Lemuria* (London: The Theosophical Publishing House, Ltd., 1925), p. 11. (This book was originally published in 1904.)
4. Ibid., p. 8.
5. Ibid., p. 20.
6. Ibid., p. 20.
7. Ibid., p. 24.
8. Joseph Campbell, *The Masks of God: Primitive Mythology* (New York: The Viking Press, Inc., 1959), p. 441.
9. Sillar and Meyer, *The Symbolic Pig*, p. 11.
10. John Bradbury, *Travels in the Interior of America in the Years 1809, 1810 and 1811 (2nd Ed.)* (London: Sherwood, Neely and Jones, 1819), p. 63.
11. Charles Towne and Edward Wentworth, *Pigs from Cave to Cornbelt* (Norman, Oklahoma: University of Oklahoma Press, 1949), p. 63.
12. Ibid., p. 61.
13. William Youatt and W. C. L. Martin, *The Hog* (New York: Orange Judd & Co., 1855), p. 33.
14. Joseph Farqua, *The Mythology of Western Civilization: From Ancient Greece to Arkansas* (New York: Doubleday & Co., Inc., 1964), p. 982. As is widely known, Mr. Farqua is an established authority on ancient history, myth and domestic bugs. The variety of his interests is due in part to his having spent the entire Second World War shipwrecked on a Pacific island with a complete (or nearly complete) set of the Encyclopaedia Britannica. His one big gap in knowledge consists in all the things that fall within the category of BER–CROG (Volume 5). For example, when once asked to compare the mating habits of the arctic snow owl to the spawning patterns of the Yucatán crawdad, he could only respond, "The arctic snow owl mates during the early spring months and lays its eggs curiously on the frozen surfaces of the tundra, and unlike the owls in other regions of the earth, feeds its young on dead lemmings." (Obviously an incomplete answer, due, no doubt, to his failure ever to compensate for not having had access to Volume 5.)

15. John Hunt, *A World Full of Animals* (New York: David McKay Co., Inc., 1969), p. 331.
16. Arthur Schopenhauer, *The Will in Nature* (New York: Doubleday & Co., Inc., 1902), p. 82.

CHAPTER II

1. John Hunt, *A World Full of Animals* (New York: David McKay Co., Inc., 1969), p. 250.
2. The Larousse Encyclopedia of Animal Life (London: Paul Hamlyn Ltd. McGraw-Hill Book Co., 1967), p. 589.
3. Joseph Campbell, *The Masks of God: Primitive Mythology* (New York: The Viking Press, Inc., 1959), p. 361.
4. Charles Towne and Edward Wentworth, *Pigs from Cave to Cornbelt* (Norman, Oklahoma: University of Oklahoma Press, 1949), p. 55.
5. Virgil, *The Aeneid.* Translation by John Dryden appearing in *Complete Poetical Works of Dryden,* edited by George R. Noyes (Boston: Houghton Mifflin Co.), p. 625.
6. "The Story of the Sheep and Hog," (*The Prairie Farmer,* Vol. 101, No. 10, March 9, 1929), p. 7.
7. F. C. Sillar and R. M. Meyer, *The Symbolic Pig* (London: Oliver and Boyd, 1961), p. 12.
8. Ibid., p. 6.
9. Herbert R. Datrum, *The U. S. Hog: His Life and Times* (Little Rock: Armadillo Press, 1945), p. 118.
10. Bartolomé de las Casas, *Historia de las Indias, Book I* (Madrid: Imprenta de Miguel Ginesta, 1876), p. 3.
11. Towne and Wentworth, *Pigs from Cave to Cornbelt,* p. 82.
12. Ibid., p. 88.
13. Richard Parkinson, *A Tour in America in 1798, 1799 and 1800* (London: J. Harding, 1805), p. 290.
14. William Faux, *Memorable Days in America: Being a Journal of a Tour to the United States* (London: W. Simpkin and R. Marshall, 1823), p. 91.
15. Thomas F. DeVoe, *The Market Assistant* (Boston: Houghton Mifflin Co., Inc., 1867), p. 483.

CHAPTER III

1. The Larousse Encyclopedia of Animal Life (London: Paul Hamlyn Ltd. McGraw-Hill Book Co., 1967), p. 586.
2. Charles Godfrey Leland, *Gypsy Sorcery and Fortune Telling* (New York: Dover Publications, Inc., 1971), p. 85.
3. Ibid., p. 89.
4. Ibid., p. 95.

CHAPTER IV

1. F. C. Sillar and R. M. Meyer, *The Symbolic Pig* (London: Oliver and Boyd, 1961), p. 97.
2. William Youatt and W. C. L. Martin, *The Hog* (New York: Orange Judd & Co., 1855), p. 24.
3. Charles Darwin, *The Descent of Man* (London: John Murray, 1870), p. 545.

CHAPTER V

1. John Hunt, *A World Full of Animals* (New York: David McKay Co., Inc., 1969), p. 252.
2. Joseph Strutt, *The Sports and Pastimes of the People of England* (London: J. White, 1845), p. 187.
3. F. C. Sillar and R. M. Meyer, *The Symbolic Pig* (London: Oliver and Boyd, 1961), p. 26.
4. Charles Towne and Edward Wentworth, *Pigs from Cave to Cornbelt* (Norman, Oklahoma: University of Oklahoma Press, 1949), p. 68.
5. William Youatt and W. C. L. Martin, *The Hog* (New York: Orange Judd & Co., 1855), p. 49.
6. Towne and Wentworth, *Pigs from Cave to Cornbelt*, p. 37.
7. Youatt and Martin, *The Hog*, p. 55.
8. David Low, *On the Domesticated Animals of the British Isles* (London: Longmans, Brown, Green and Longmans, 1853), p. 403.
9. Joseph Farqua, *The Ice Age Boar: Truth or Maybe* (New York: Doubleday & Co., Inc., 1941), p. 1283.

CHAPTER VII

1. .William Youatt and W. C. L. Martin, *The Hog* (New York: Orange Judd & Co., 1855), p. 45.
2. Ibid., p. 38.
3. Ibid., p. 108.
4. F. C. Sillar and R. M. Meyer, *The Symbolic Pig* (London: Oliver and Boyd, 1961), p. 45.
5. Ibid., p. 44.
6. Rev. W. B. Daniel, *Rural Sports* (London: Logley and Goatworth, 1803), p. 111.
7. Youatt and Martin, *The Hog*, pp. 36–37.
8. Ibid., p. 34.

CHAPTER VIII

1. F. C. Sillar and R. M. Meyer, *The Symbolic Pig* (London: Oliver and Boyd, 1961), p. 115.
2. Joseph Campbell, *The Masks of God: Occidental Mythology* (New York: The Viking Press, Inc., 1964), pp. 17–19.
3. Sir James G. Frazer, *The Golden Bough: A Study in Magic and Religion* (New York: The Macmillan Co., Paperbacks Edition, 1960), pp. 543–45.
4. Ibid., p. 547.
5. Joseph Campbell, *The Masks of God: Primitive Mythology* (New York: The Viking Press, Inc., 1959), p. 432.
6. Joseph Campbell, *The Masks of God: Creative Mythology* (New York: The Viking Press, Inc., 1968), p. 206.
7. Campbell, *The Masks of God: Occidental Mythology*, p. 299.
8. Campbell, *The Masks of God: Primitive Mythology*, p. 446.
9. Sillar and Meyer, *The Symbolic Pig*, p. 16.
10. Frazer, *The Golden Bough: A Study in Magic and Religion*, p. 547.

CHAPTER IX

1. Gordon W. Allport, *The Nature of Prejudice* (Garden City, New York: Doubleday & Co., Inc., 1954), p. 11.
2. Alfred North Whitehead, "Uses of Symbolism," *Symbolism in Religion and Literature,* ed. Rollo May (New York: George Braziller, Inc., 1961), p. 250.

Bibliography

AGEE, JAMES. *Letters of James Agee to Father Flye.* New York: George Braziller, 1962.

ALLPORT, GORDON W. *The Nature of Prejudice.* New York: Doubleday, 1954.

BRADBURY, JOHN. *Travels in the Interior of America in the Years 1809, 1810 and 1811 (2nd Ed.).* London: Sherwood, Neely and Jones, 1819.

BROWN, NORMAN O. *Life Against Death.* New York: Vintage Books, 1959.

CAMPBELL, JOSEPH. *The Masks of God: Creative Mythology.* New York: The Viking Press, 1968.

———. *The Masks of God: Occidental Mythology.* New York: The Viking Press, 1964.

———. *The Masks of God: Primitive Mythology.* New York: The Viking Press, 1959.

DANIEL, REV. W. B. *Rural Sports.* London: Logley and Goatworth, 1803.

DATRUM, HERBERT R. *The U. S. Hog: His Life and Times.* Little Rock: Armadillo Press, 1945.

DE LAS CASAS, BARTOLOMÉ. *Historia de las Indias, Book I.* Madrid: Imprenta de Miguel Ginesta, 1876.

DEVOE, THOMAS F. *The Market Assistant.* Boston: Houghton Mifflin, 1887.

ELLIOT, W. SCOTT. *The Story of Atlantis and the Lost Lemuria.* London: The Theosophical Publishing House, 1925.

FARQUA, JOSEPH. *The Ice Age Boar: Truth or Maybe.* New York: Doubleday, 1941.

———. *The Mythology of Western Civilization: From Ancient Greece to Arkansas.* New York: Doubleday, 1964.

FAUX, WILLIAM. *Memorable Days in America: Being a Journal of a Tour to the United States.* London: W. Simpkin and R. Marshall, 1823.

HUNT, JOHN. *A World Full of Animals.* New York: David McKay, 1969.

The Larousse Encyclopedia of Animal Life. London: Paul Hamlyn Ltd., 1967.

LELAND, CHARLES GODFREY. *Gypsy Sorcery and Fortune Telling.* New York: Dover Publications, 1971.

LOW, DAVID. *On the Domesticated Animals of the British Isles.* London: Longmans, Brown, Green and Longmans, 1853.

O'SULLIVAN, SEAN. *Folktales of Ireland.* Chicago: The University of Chicago Press, 1966.

PARKINSON, RICHARD. *A Tour in America in 1798, 1799 and 1800.* London: J. Harding, 1805.

PRINCE MAXMILLIAN of Weid. *Travels in the Interior of North America.* London: Ackerman and Co., 1843.

Ralston Purina Co. *Checkerboard Serviceman's Pocket Manual.* St. Louis: The Ralston Purina Company, 1963.

RANZBOTTOM, DR. E. L. *When Will Civilization Stop Fooling Around and Get Serious for a Change?* Toronto: Moose Press, 1971.

SCHOPENHAUER, ARTHUR. *The Will in Nature.* New York: Doubleday, 1902.

SILLAR, F. C., and MEYER, R. M. *The Symbolic Pig.* London: Oliver and Boyd, 1961.

STRUTT, JOSEPH. *The Sports and Pastimes of the People of England.* London: J. White, 1845.

TAYLOR, JAMES. *A Porcine History of Philosophy and Religion.* Nashville: Abingdon Press, 1972.

"The Story of the Sheep and Hog," *The Prairie Farmer,* Vol. 101, No. 10, 1929, 7–10.

TOWNE, CHARLES, and WENTWORTH, EDWARD. *Pigs from Cave to Cornbelt.* Norman, Oklahoma: University of Oklahoma Press, 1949.

VALDEZ, "POSSUM" R. *The Hog—Take It or Leave It.* New Orleans: Gumbo Press, 1955.

VIRGIL. *The Aeneid.* Translation by John Dryden appearing in *Complete Poetical Works of Dryden,* edited by George R. Noyes. Boston: Houghton Mifflin Co.

WHITEHEAD, ALFRED NORTH. "Uses of Symbolism," *Symbolism in Religion and Literature,* ed. Rollo May. New York: George Braziller, 1961.

YOUATT, WILLIAM, and MARTIN, W. C. L. *The Hog.* New York: Orange Judd & Co., 1855.

Hog Potential Movement

O DULCE PORCUS

_____
Member's Name

Clip along dotted line, laminate and put in your wallet. Congratulations.